Environmental Remediation Technologies, Regulations and Safety Series

Environmentalism: Environmental Strategies and Environmental Sustainability

ENVIRONMENTAL REMEDIATION TECHNOLOGIES, REGULATIONS AND SAFETY SERIES

Contaminated Soils: Environmental Impact, Disposal and Treatment
Robert V. Steinberg (Editor)
2009. ISBN: 978-1-60741-791-0

Environmentalism: Environmental Strategies and Environmental Sustainability
Elias Kahraman and Ahmed Baig (Editors)
2010. ISBN: 978-1-60741-634-0

ENVIRONMENTAL REMEDIATION TECHNOLOGIES, REGULATIONS AND SAFETY SERIES

ENVIRONMENTALISM: ENVIRONMENTAL STRATEGIES AND ENVIRONMENTAL SUSTAINABILITY

ELIAS KAHRAMAN
AND
AHMED BAIG
EDITORS

Nova Science Publishers, Inc.
New York

For permission to use material from this book please contact us:
Telephone 631-231-7269; Fax 631-231-8175
Web Site: http://www.novapublishers.com

LIBRARY OF CONGRESS CATALOGING-IN-PUBLICATION DATA

Kahraman, Elias.
Environmentalism : environmental strategies and environmental sustainability / Elias Kahraman and Ahmed Baig.
p. cm.
Includes bibliographical references and index.
ISBN 978-1-60741-634-0 (hardcover : alk. paper)
1. Environmentalism. I. Title.
GE195.K34 2009
304.2--dc22

2009026861

Published by Nova Science Publishers, Inc. ✦ New York

CONTENTS

PREFACE

Environmentalism is a broad philosophy and social movement centered on a concern for the conservation and improvement of the environment. This book puts forward some key strategies for promoting Cleaner Production in China, for instance, integrating CP into sustainability strategies, technology innovations and industrial ecology. Furthermore, the authors examine the Energy Masting Planning, a comprehensive plan that addresses energy supply and consumption through 2020. The plan includes energy efficiency, renewable energy and infrastructure and land use policies and emphasizes both the benefits and the limits of the approach. Furthermore, removal of toxic and heavy metal contaminants from aqueous environments is one of the most important environmental issues to face the world. In this book, aerobic degradation through bioaccumulation by bacteria and microalgae and enzyme-catalyzed reduction-based remediation of toxicants from waste waters are discussed. Other chapters in this book examine the attitudes of university students towards the environment and environmental problems, the influence on the causes of forest decline and an analysis of specific factors that influence the nominal median price of single-family homes across states, with a particular emphasis placed on the capitalization of environmental factors such as environmental pollution in the form of toxic chemical releases.

Chapter 1 – The explosion in biological knowledge is touching almost all areas of science and engineering. Improved understanding can lead to improved design, operation, and maintenance of environmental engineered systems. The integration of molecular biology tools in environmental engineering is a challenge. The authors discuss our views on the following two critical issues: The opening section provides an informative introduction to Environmental Biotechnology. Like white biotechnology, environmental biotechnology, often referred to as "grey biotechnology", also focuses on sustainability. For instance, environmental biotechnology deals with the treatment of sewage water, the purification of exhaust gas or the decontamination of soils or ground water using specific microorganisms. Environmental technologies are likely to play a more important role as they involve the use of environment-friendly mechanisms. This means starting environmental protection from the very beginning of production. In 1992, at the 21st Agenda of the UN Conference on Environment and Development in Rio de Janeiro, environmental biotechnology was, for the first time, seen to have an important role to play in tackling environmental problems. Traditional biotechnological processes are an integral part of classical "end of pipe" technologies. One could also refer to this as "first generation environmental biotechnology". Traditional environmental biotechnology therefore relies on the metabolic by-products of naturally occurring organisms. The rapid progress in biotechnology, molecular biology and gene technology broadens the application spectrum of environmental biotechnology: Desired metabolic products can be provided in a targeted and inexpensive way. Apart from genetically optimized organisms, isolated enzymes or biological products are available in nearly all amounts desired. Second generation environmental biotechnology is more and more based on high-tech methods and processes. This section provides a critical review of recent biotechnological approaches to environment protection. The second section focuses on the application of biotechnology in the bioremediation of environment pollution. Bioremediation is an alternative to traditional remediation technologies such as landfilling or incineration. Bioremediation works by either transforming or degrading contaminants to non-hazardous or less hazardous chemicals. A number of bioremediation strategies have been developed to treat contaminated wastes and sites. Several strategies are currently being developed on biological wastewater treatment including aquatic plants, membrane and microorganisms. Bioremediation is also applied in removing pollutants from contaminated soil and petroleum industry. Bioremediation has become one of the most rapidly developing fields of environmental restoration.

Chapter 2 – This chapter defines the concept of energy master planning and emphasizes both the benefits and the limits of the approach. It then establishes the importance of leadership and institutionalization of the plan to set the groundwork for successfully achieving the goals. This includes identifying and engaging key stakeholders in the process of developing the plan. Finally, the use of scenario modeling illustrates how alternative policy proposals are vetted for their relative impact on macroeconomic and environmental indicators for the state. This quantitative analysis—though far from conclusive—provides a level of legitimacy to the planning process and should lessen potential critiques. Involving stakeholders and experts in modeling is as important as engaging them in the development of the policy platform.

Chapter 3 – Removal of toxic and heavy metal contaminants from aqueous environments is one of the most important environmental issues to face the world. Conventional treatment technologies (chemical precipitation, ion-exchange resins, membrane separation, etc.) for removal of toxic substances from aqueous solution are not economical and also generate huge quantities of toxic chemical sludge. Hence, research has been intensified on the search for alternative and innovative wastewater treatment techniques. The techniques mostly have focused on microorganisms. Metabolically active biomasses and more complex systems such as employing enzymes are considered to be potential instruments for removal of these pollutants from waste effluents. Attention has been paid recently to the enzyme technology. A large number of enzymes from a variety of different microorganisms have been reported to play an important role in an array of waste treatment applications. However there are challenges to overcome in these systems due to sensitivity to extremes of pH, high concentration of pollutants, and requirement of external metabolic energy. Hence the isolation and selection of resistant strains are crucial in order to overcome the limitations of employing living cells. Studies on bioaccumulation have shown that some strains of bacteria, algae (microalgae and cyanobacteria) and fungi isolated from contaminated sites possess excellent to survive capability and have high tolerance to various toxic metals. These microorganisms may be promising candidates for detoxifying and removing toxicants from wastes.

In this review aerobic degradation through bioaccumulation by bacteria and microalgae and enzyme-catalyzed reduction-based remediation of toxicants from waste waters are discussed.

Chapter 4 – In recent years, the relevant national ministries and departments and media have increased the publicity of environmentalism and its ideas. Because of environmental deterioration and its negative consequences, Environmental respect has become a need rather than just an idea. But, the authors have verified that market forces are insufficient to regulate the impact of companies. For this reason, it is necessary to articulate a set of norms that consider the peculiarities of each industry and force companies to minimize the environmental impact of their activities. Efficient environmental protection requires global, as well as regional ecological strategies. During the last decade numerous attempts at regional and global environmental protection have been made. But industrial environmental protection is still, above all, a national matter. For China, which is now in the period of accelerating industrialization and urbanization, the per capita resource amount is below than the world average. So the circular economy will become an important practice for the country to sustain its socio-economic development. And according to the China Council for International Cooperation on Environmental Protection and Sustainable Development Task Force on Circular economy and Cleaner Production, Cleaner Production is the cornerstone of Circular economy. If China wants to vigorously build a resource-saving and environmental-friendly society, Cleaner Production should be an essential part of any comprehensive pollution management system, at the enterprise or the national level. This article puts forward some key strategies for promoting Cleaner Production in China, for instance, integrating CP into sustainability strategies, technology innovations, and industrial ecology. Lack of capacity to track and evaluate the effects of CP programmes, and to disseminate this information, still seems to be one of key challenges for CP initiatives. So development of a Cleaner Production indicator system and recommendation of technical criteria is essential. How we design, how we meet people's needs, how we use products, and leading to excessive consumption are some of the challenges we face to reach a sustainable and sustaining future. The efficient solution will require social, economic and technological innovations. In this article, Economic Policies for Sustained Development is described, and advocated as an important first step in a strategy moving towards economic-cum-environmental sustainability. Drastic changes and bold initiatives are suggested and required to establish true sustainable development for China.

Chapter 5 – Motor development research has experienced greater frequency in top-tier science journals over the past 25 years. While some dialogue has been of *environmental factors*, to a much greater extent, discussions of motor development have focused on the biological, psychological, cognitive, or movement aspects of change. Few would disagree that to truly understand the mechanisms and processes associated with change in human form and level of motor behavior, a more comprehensive and integrated approach is needed – one that links biological and environmental theory. One such approach and the focus of this chapter is the *Developmental Systems Perspective*.

Chapter 6 – A meso-scale pilot plant was set up to test the efficiency of a phyto-bioremediation technique applied to slightly contaminated marine sediments dredged from the port of Livorno (central Italy). Different plants were used in combination (a grass, *Paspalum vaginatum* sp. and a shrub, *Tamarix gallica* sp.) with earthworms (*Eisenia foetida* sp.) to improve the bio-physical properties of the contaminated sediments. The objective was to transform the degraded sediment into a fertile and nutrient-rich matrix, that can be at the same time efficiently phytoremediated. The plant-microorganism-earthworm association is a powerful ecological system working at the same time, at three levels: plants release organic exudates, microorganisms mineralize them and grow so releasing nutrients for plants, and earthworms feed on microorganisms. The experiment was performed in columns at pilot-scale. The sediment was biophysically preconditioned with manure and inert bulking materials (5:1 v/v). This constructed system allowed plants to grow in an initial fine textured and salty medium, unfriendly for plant species. The efficiency of the recreated ecosystem was confirmed by the improvement of physical properties of the sediment. This included both permeability and water content, starting from an initial value of 10^{-5} m sec^{-1} and 60% (asphyxial medium) respectively, and reaching at the end of the experimentation, value of 10^{-3} m sec^{-1} and 40%, respectively. Furthermore, significant stimulation of ecological parameters that drive the bioremediation were obtained, such as nutrients (i.e. carbon, nitrogen, phosphorous), microbial dehydrogenase and β-glucosidase enzymes, and humic extractable carbon. Stepwise regression statistics was used to focus the correlation between the physical and biochemical properties of the reconstructed matrix. The statistical approach confirmed that the presence of organic macro-aggregates in the sediment (e.g. humic substances) allows the permeability of the medium to increase, also stimulating enzyme activities (e.g. dheydrogenase and β-glucosidase) involved in the mineralization of the carbon-cycling and mineral nutrient release.

Chapter 7 – This chapter was designed and conducted to determine the attitudes of university students towards the environment and environmental problems.

In the scope of this descriptive study, a questionnaire form was prepared by using an environmental attitude scale to collect study data. This scale was applied between 30 September 2008 and– 8 October 2008 to students in the Department of Family and Consumer Sciences, Faculty of Industrial Arts Education, Gazi University (n: 358). Three hundred students completed in and returned the questionnaire. The SPSS 13.0 statistical program was used to study data. Mean value, standard deviation, T-test, Levene test, Tamhane test and LSD test were used in the analysis.

The authors found out that there was no statistically meaningful difference between the mean attitude scores of groups when controlled for variables of age, sex, number of siblings, any environment education taken before university, any out-of-school environment training, and the responsibility of individuals, state and conservationist institutions in the solution of environmental problems. However, students' opinions about how they should be assessed in the ecology class differed significantly in terms grade and income variables. Students listed air and water pollution (42.0%), population growth and urbanization (39.3%), and deforestation (18.6%) as the three most important environmental problems. Considering students' total "Environmental Attitude Scale" score, the highest mean attitude score was given to the "air, water and soil are inexhaustible resources" item while the lowest mean score was given to the "thinning ozone layer threatens all humanity" item.

Chapter 8 – The authors examined the influence of two causes of forest decline, i.e., forest fire and pine wilt disease, on discharge duration curves in a paired-catchment experiment in the Tatsunokuchi-yama Experimental Forest, Japan (34°42'N, 133°58'E). Regression lines between discharge duration curves for the Kita-tani and Minami-tani catchments were calculated during the control period (36 years). The observed discharge was then compared to that estimated from the regression lines for the disturbed period (19 years). Disturbance generally increased daily discharge, except on several dozen days. The observed discharge was increased by approximately 15%, 35%, and 43% for days 1–40, 140–320, and more than 320, respectively, in the discharge duration curve.

Further, from a comparison of the discharge duration curves of the two catchments, the authors developed a new index for estimating the short-term influence of small forest disturbances on discharge rates in the Minami-tani catchment. The authors used this index to estimate the influence of disturbance, and the recovery speed from that disturbance, in a 22.61-ha catchment.

Chapter 9 – Using a semi-log model specification, this empirical chapter investigates determinants of nominal median single-family home price differentials for the year 2005 using state-level data. Although the literature on geographic living-cost differentials is well developed, the literature on geographic housing price differentials is much less so. Given the relatively large impact of housing prices on overall living costs, this research seeks to address this issue and shed light on specific factors influencing the nominal median price of single-family homes across states. Particular emphasis is placed on the capitalization of environmental factors such as environmental pollution in the form of toxic chemical releases and favorable climatic conditions into housing prices. As control variables, several economic variables and a measure of fiscal surplus are also included in the analysis. The semi-log results imply that the natural log of the nominal median price of single-family homes is negatively a function of toxic chemical releases. In addition, it is positively a function of measures of both warm climate and sunny climate. Furthermore, the natural log of the median price of single-family homes is an increasing function of median family income, the percentage employment growth rate, the extent of unionization, and the magnitude of fiscal surplus.

Chapter 10 – In this chapter, the demographic dimension of climate change is discussed. In the first part of the chapter, I analyze the leverage of demographic dynamics on demand of energy-intensive consumption goods and carbon dioxide emissions. Several demographic factors are related to greenhouse gas emissions. Population growth is the most obvious one: the larger the population the larger the amount of emissions, given fairly constant per-capita carbon emissions over time. Population age structure is another important factor: given age-specific consumption profiles of energy-intensive goods, changes in population age structure have a considerable impact on overall levels of consumption and greenhouse gas emissions. Household size affects the extent of economies of scale in energy use from cohabitation: larger households are associated with lower per capita energy consumption. In the second part of the chapter, I discuss the role of the economic structure in amplifying or softening the leverage of demographic dynamics on greenhouse gas emissions. Demographic agents are considered in their role of consumers and producers: the implications of a consumption-based versus a production-based approach to the evaluation of the effect of demographic dynamics on greenhouse gas emissions are explicated. The hypothesis of existence of environmental Kuznets curve is analyzed and its validity in the context of greenhouse gas emissions is discussed. The main point of the chapter is that demographic quantities have an important leverage on greenhouse gas emissions. Their effect is mediated by economic and technological factors. A comprehensive policy approach to address the issue of climate change, that takes into account both demographic and economic dynamics is suggested.

In: Environmentalism: Environmental Strategies… ISBN: 978-1-60741-634-0
Editors: E. Kahraman and A. Baig, pp.1-40

Chapter 1

DEVELOPMENTS IN APPLYING BIOTECHNOLOGY TOOLS IN ENVIRONMENTAL ENGINEERING

Bing-Gang Ma*, Xiao-Yan Lu, Zhi-Ping Wu and Hu-Ping Zhang

Key Oasis Eco-agriculture Laboratory of Xinjiang Production and Construction Group, Department of Horticulture, Agricultural College, Shihezi University, Shihezi 832000, People's Republic of China

ABSTRACT

The explosion in biological knowledge is touching almost all areas of science and engineering. Improved understanding can lead to improved design, operation, and maintenance of environmental engineered systems. The integration of molecular biology tools in environmental engineering is a challenge. We discuss our views on the following two critical issues: The opening section provides an informative introduction to Environmental Biotechnology. Like white biotechnology, environmental biotechnology, often referred to as "grey biotechnology", also focuses on sustainability. For instance, environmental biotechnology deals with the treatment of sewage water, the purification of exhaust gas or the decontamination of soils or ground water using specific microorganisms. Environmental technologies are likely to play a more important role as they involve the use of environment-friendly mechanisms. This means starting environmental protection from the very beginning of production. In 1992, at the 21st Agenda of the UN Conference on Environment and Development in Rio de Janeiro, environmental biotechnology was, for the first time, seen to have an important role to play in tackling environmental problems. Traditional biotechnological processes are an integral part of classical "end of pipe" technologies. One could also refer to this as "first generation environmental biotechnology". Traditional environmental biotechnology therefore relies on the metabolic by-products of naturally occurring organisms. The rapid progress in biotechnology, molecular biology and gene technology broadens the application spectrum of environmental biotechnology: Desired metabolic products can be provided in a targeted and inexpensive way. Apart from genetically optimized organisms, isolated enzymes or biological products are available in nearly all amounts desired.

* E-mail: mbg_agr@shzu.edu.cn

Second generation environmental biotechnology is more and more based on high-tech methods and processes. This section provides a critical review of recent biotechnological approaches to environment protection. The second section focuses on the application of biotechnology in the bioremediation of environment pollution. Bioremediation is an alternative to traditional remediation technologies such as landfilling or incineration. Bioremediation works by either transforming or degrading contaminants to non-hazardous or less hazardous chemicals. A number of bioremediation strategies have been developed to treat contaminated wastes and sites. Several strategies are currently being developed on biological wastewater treatment including aquatic plants, membrane and microorganisms. Bioremediation is also applied in removing pollutants from contaminated soil and petroleum industry. Bioremediation has become one of the most rapidly developing fields of environmental restoration.

INTRODUCTION

This is a century of biotechnology. Many of the scientific breakthroughs we predicted more than twenty years ago are now coming out of the laboratory and into widespread commercial use, bringing us face to face for the first time with the promises and perils of this new era. Although GM technology was first developed for plants in 1983, the First GM crop (termed more appropriately as 'biotech crop') reached the markets during the mid-1990s and during the past decade the development and adoption of transgenic technology has progressed rapidly. In 2007, the global area of biotech crops increased for the twelfth consecutive year at an annual growth rate of 12%. While the technology was initially applied in developed countries, in 2007 biotech crops were grown by 12 million farmers in 23 countries covering 114.3 million hectares and over 90% of the beneficially farmers were resource-poor in developing countries, with increased incomes from biotech crops contributing to the alleviation of poverty (James, 2007). Significant progress has also been made towards the elucidation of plant genes that, as single variables, are able to improve yield (Camp, 2005). Much has been written about the various, different perceptions of transgenic crops. To some they represent an exciting opportunity to use modern biological techniques to improve the quality of agricultural plants and their products, whilst to others they are seen as an insidious attempt to extend the power of multinational companies and thereby threaten the well being of subsistence farmers, as well as the ecology of fragile ecosystems throughout the world. It is generally agreed that this rise of opposition is contextual; in particular the commercialization of GM products in Europe arrived in an adverse context as it followed the occurrence of two major health crises, namely contaminated blood (France), and mad cow disease outbreaks (UK) (Joly and Lemarié, 1998). In addition, the rise of opposition is linked to a lack of public understanding of current agriculture. There are two perceptions: (1) 'Traditional' agriculture is well perceived (often associated with the preservation of nature, quality products, and small farming systems). (2) 'Modern' agriculture is considered to have had a negative impact on all of the above factors. Therefore, the use of GMOs in agriculture is associated with modern agriculture and, thus, motivates a negative perception.

1. Environmental Biotechnology

Biotechnology could be widely applied in the technological development of environmental protection. Environmental biotechnology is newly developing technology in the 21st century. It is the sector of biotechnology that concerns microbial processes for efficient, effective and economic waste treatment, particularly the elimination of noxious pollutants; microbial processes for the remediation of previously polluted natural environments and for cleaning up dirty naturally occurring feedstock; the systematic evaluation and control of microbially mediated degradation processes in engineered environments, and the impact of both intentional and accidental contamination of natural environments with non-indigenous microbes. It is generally acknowledged that environmental pollution, resource problem, ecology destroy, and health problem and so on can be solved by environmental biotechnology. The term environmental biotechnology has a certain air of modernity when in fact it has a long history of use, if one considers the underlying principles and not the appellation. However, as part of its complex meaning, there is a dynamic new definition and purpose in this discipline with regard to bioremediation. The ability to probe the environment at the molecular level with exquisite methods, to create a new awareness of fundamental biological processes therein, has created an important new paradigm in remediation engineering design and management. Further, biological lines of evidence made extremely robust through the merger of biotechnology and environmental science are poised to be incorporated into the very fabric of site evaluation and disposition at the regulatory level. At the operational level, the field of environmental biotechnology is driven by the " omics," the common suffix for disciplines like genomics, proteomics, and metabolomics. Environmental biotechnology has made great progress in the following fields: Monitoring technology and treatment of dioxin-like chemicals and cyanobacterial toxins, biofiltration for air pollution control, solid waste treatment and reutilization, bioremediation of soil pollution, risk assessment and control of endocrine disturbing substances in environment, wastewater treatment, clean production and recycling economy.

2. Gene Engineering and Environmental Safety

Plant biotechnology is founded on the principles of cellular totipotency and genetic transformation. Many experimental strategies that led to the production and commercialization of biotech (transformed or transgenic) plants expressing many useful genes, and emphasizes the beneficial effects of plant biotechnology on food security, human health, the environment, and conservation of biodiversity. In the last decade, the technique to genetically modify crop plants has gained more and more interest in terms of bioproduction of heterologous proteins. Plants have been discovered as a possible source for large amounts of cost effective recombinant protein. Main application fields are therapeutics for use in animal and human health, diagnostics, and technical enzymes.

Potential impacts of genetically modified organisms (GMOs) on the environment have been addressed in several authoritative reviews (Rissler and Mellon, 1996; Traynor and Westwood, 1999; Wolfenbarger and Phifer, 2000; Letourneau and Burrows, 2002). Evidence from recent studies is then presented with the aim of suggesting possible improvements to contemporary risk assessment.

2.1. Transgenic Biotechnology

Development of transgenic plants in an effective manner is important pre-requisites for sustainable and economic use of biotechnology for crop improvement (Sharma et al., 2002). The process of genetic transformation involves several distinct stages, namely insertion, integration, expression and inheritance of new DNA and, in most current transformation systems, a tissue culture stage that is prerequisite for the ultimate recovery of transgenic plants. Although almost every significant crop species has been successfully transformed, in many species the development of rapid, highly efficient, and routine transformation systems is still in progress and thus represents a bottleneck in the development of stable high yielding transgenic plants (Sharma et al., 2002). During the last two decades the development and adoption of transgenic technology has progressed rapidly. Although almost every significant crop species has been successfully transformed, the absence of a rapid, highly efficient, and routine transformation system in many crop species represents a problem in the development of stable high yielding transgenic plants. Almost every type of plant tissue has been subjected to gene transformation experiments and several methods for transformation have been developed. However, only *Agrobacterium* mediated transformation and particle bombardment are being used routinely for successful development of transgenic plants and most of the efforts are focused on herbicide and insect tolerance. The main drawback of the *Agrobacterium*-mediated system is its dependence on tissue culture regeneration ability of a plant species or a variety of crop plant and the transformation ability of that species or variety by *Agrobacterium*. Although particle bombardment circumvents the problems associated with agro-infection, regeneration of a complete plant still requires tissue culture. For this reason, there is considerable interest in developing plant transformation methods that circumvent tissue culture steps. Gene targeting is still an elusive goal for the plant biotechnologist. It will continue to be an interesting area for research in the years to come because of its outstanding potential. Until an efficient method for creating precise changes at the DNA level is developed, RNAi will continue to be the major tools for gene silencing and the study of gene functions in plants. (Akhond and Machray, 2008)

Crucial steps in gene targeting in plants include: the design of the gene targeting construct, the choice of the tissue to be transformed, the selection of transformants and their regeneration, and the screen for clean, targeted transformants. In the reported rice GT strategy, a combination of all these factors probably contributed to the reported success (Hohn and Puchta, 2003). A number of different approaches have been taken to combine or 'pyramid' transgenes in one plant and have met with varying degrees of success. These approaches include sexual crossing, re-transformation, co-transformation and the use of linked transgenes. These methods have been utilized to generate the 'stacked trait' varieties now increasingly being grown in the USA. In the future it is likely that novel, alternative technologies will be developed that use single transgenes to coordinate the expression of multiple genes. Amongst the constructs being tested at present are some modelled on viral polyproteins can be used to simultaneously introduce multiple protein-coding genes into plants.

2.2. Development in Transgenic Plants

2.2.1. The Development of Transgenic Plants

The various types of research programme will be categorised according to the various objectives which range from industrial applications, to alterations in seed quality traits and food constituents, to more long term strategies (Dunwell, 1999) to use crops as a source of valuable pharmaceutical products.

During the past decade the development and adoption of transgenic technology has progressed rapidly. In 2007, biotech crops were grown by 12 million farmers in 23 countries covering 114.3 million hectares. This progress can be attributed to development in molecular genetics, plant transformation and regeneration techniques and a better understanding of the underlying processes involved in DNA recombination. While almost every significant crop species has been successfully transformed, in many species the development of rapid, highly efficient, and routine transformation systems is still in progress.

The 'first generation' products principally comprise soybean, maize (corn) and cotton resistant to herbicides and/or insects. The wide range of 'second generation' products, under development and testing in many commercial and academic laboratories, include examples for a variety of food, medical, veterinary and industrial purposes. Most of the GM crop varieties has around the world contain introduced microbial genes providing tolerance to broad spectrum herbicides (e.g., glufosinate or glyphosate) and/or resistance to insect pests such as the European corn borer. These varieties now predominate in some areas of the USA, although they are still only undergoing relatively small scale testing in Europe, where the need for GM products and their safety are the subject of concerted opposition from some individuals and groups. Most participants in this debate accept that the lack of perceptible consumer benefit has been a significant deterrent to acceptance of this technology, particularly in the context of other food scares. Whether self interest will ever overcome this resistance may depend partly on the development of new products where the real and obvious benefit can be seen to be outweigh the assumed and non-obvious risk.

2.2.2. Applications of Transgenic Plants

More food and feed have to be grown on less land. Fortunately, in the 1970s, new tools for more efficient plant breeding arose, one of which is biotechnology. As agrochemistry and mechanisation have reached a level where further progress is neutral in relation to plant production—scientific progress helps environmental needs—only breeding may close the yield gap by a combined application of classical and molecular techniques (Gerhard Wenzel, 2006). Application of molecular plant breeding is now contributing to discoveries of genes and their functions opening new avenues for basic plant biology research (Moose and Mumm, 2008). Transgenes could also be transferred using much cheaper conventional plant breeding techniques to locally adapted crop varieties in countries with less sophisticated research facilities. The combination of traditional crop improvement techniques and modern biotechnological techniques can contribute substantially to human well-being by providing renewable sources of food, feed, chemicals, pharmaceuticals, energy and for the creation of a sustainable environment for the developed as well as for the developing countries of the world.

One of the potential environmental benefits of transgenic plants is their use in bioremediation, the process of decontamination of polluted soils or water. For example, mercury pollution is a major environmental problem accompanying industrial activities, with most of the mercury released being retained in the soil as complexes of the toxic ionic mercury (Hg^{2+}), which then can be converted by microbes into the even more toxic methylmercury. Mercury detoxification of the soil can also occur by microbes converting the ionic mercury into the least toxic metallic mercury (Hg0) form, which then evaporates. The remediation potential of transgenic plants carrying the *MerA* gene from *E. coli* encoding mercuric ion reductase has been evaluated in several cases. For example, a codon-optimized version of the gene was introduced into tobacco and it was found that transgenic seeds were resistant to $HgCl_2$ at 50 μM, and some of them could germinate on media containing as much as 350 μM $HgCl_2$, while the control plants were fully inhibited or died on 50 μM $HgCl_2$. The rate of elemental mercury evolution from Hg^{2+} (added as $HgCl_2$) was 5–8 times higher for transgenic plants than the control with volatilization being up to 5 times higher in the roots than in shoots or the leaves. Other compounds shown to be detoxified by transgenic plants include explosives, chlorinated solvents (Doty et al., 2000) and hydrocarbons (oil and petroleum residues). Recovery of agricultural resources (vast territories polluted with pesticides, oil, or heavy metals) was yet another important issue dealt with at the section. Bioremediation techniques, as being the least expensive and risky are to be implemented to address it. The use of rhizospheric bacteria of the genus *Pseudomonas*, which act as growth stimulators in certain plants, appears to offer the greatest promise in this respect.

One of the recurring themes concern the potential of using plants as a production system for high value compounds of medical importance. Examples include the expression of antibodies that may reduce the growth of bacteria associated with dental caries or that can be used in the treatment of human tumours. The advances in the use of plants to produce vaccines (Arntzen et al., 2001) for veterinary or medical purposes have extended greatly. Much attention to GM crops is discussion of their potential as a production system for biodegradable plastics, particularly poly (beta-hydroxybutyrate) (PHB) and poly (betahydroxyvalerate) (PHV).

2.3. Achievements of Transgenic Biotechnology

2.3.1. Improvement of Nutritional Qualities

How do genetically engineered foods compare with conventional foods in nutritional quality? This is an important issue, and one for which there will probably be much research in the future, as crops that are engineered specifically for improved nutritional quality are marketed. There are many, varied applications of transgenic technology in modifying food quality. Although some of these refer to general attributes such as resistance to freezing the majority are concerned with specific changes in the amounts of beneficial or toxic compounds.

The improvement of either the quantity or quality of seed protein is at the centre of many plant breeding programmes (Vasil et al., 2001) and there is a wide diversity of approaches to the modification of specific traits. For example, it has been claimed that expression of an inositol polyphosphate phosphatase homologue improves the total amount of protein. Amongst the novel means of modifying a quantitative trait, namely grain texture, is one comprising the introduction into wheat of genes encoding puroindoline A and B. Some projects aim to alter only a specific amino acid (e.g., methionine, lysine, tryptophan)in order to improve nutritional quality whilst other studies have more generalobjectives, These latter studies depend on the fact that the composition of highmolecular weight subunits of glutenin determines the gluten strength and influences the baking quality of bread wheat.

As well as the projects designed to modify protein quality, there are many concerned specifically with changing the profile of starch produced either in the seed or other storage organ such as the tuber of potatoes. Examples include the use of a beta-amylase gene, a rice starch granule-bound protein, a debranching enzyme, or starch synthase to produce modified starches in transgenic plants. It is also claimed that an overall increase in the level of stored carbohydrate can be achieved by overexpressing dihydro-orotase, a gene of pyrimidine metabolism.

2.3.2. Insect Resistance

Losses due to insect herbivores, estimated at 10–20% for major crops, are a significant factor in limiting food production. Engineering crop plants for endogenous resistance to insect pests has been one of the real successes of GM technology, following the commercial introduction of genetically modified maize, potato and cotton plants expressing genes encoding the entomocidal-endotoxin from *Bacillus thuringiensis* (*Bt*) in the mid-1990s (Ferry et al., 2004). In 2007, 20.3 million hectares (18%) were planted with *Bt* crops making it the third most important transgenic trait.

Although the acreage of Bt-transgenic crops has increased dramatically in recent years, there are concerns that sustained efficacy of Bt crops will be endangered by the development of resistance in populations of insect pests that they are intended to control. To date, several major pest species, including *Heliothis virescens*, *Leptinotarsa decemlineata*, and *Plodia interpunctella*, have demonstrated the ability to adapt to Bt in the laboratory. Moreover, *Plutella xylostella* (diamondback moth, DBM) has evolved sufficiently high resistance to Cry1A proteins in field settings to result in field control failures. Such evidence of insect resistance to Bt proteins emphasizes the need to develop effective resistance management strategies.

2.3.3. Herbicide Resistance

Detoxification is an important mechanism by which many plants as well as bacteria develop natural tolerance to a variety of herbicides. During the past two decades, novel and successful strategies have been developed for engineering herbicide tolerance in plants by isolating, cloning and characterizing bacterial or plant genes that detoxify specific herbicides and then transferring them to crops of choice by genetic transformation. This has made it possible to use a new generation of non-selective herbicides that are relatively inexpensive, highly potent, environmentally friendly and biodegradable, such as Roundup and Basta, by engineering the relevant resistance genes into a variety of crop species, including wheat (De Block et al., 1987; Duke, 1996; Gressel, 1996; Padgette et al., 1996; Vasil, 1996, 1999; Heck et al., 2003). The fact that nearly 75% of the biotech crops planted worldwide are engineered for herbicide resistance (mostly Roundup tolerance) is a testament to their usefulness in agriculture and their attraction to farmers.

2.3.4. Disease Resistance

Protection against diseases caused by pathogenic organisms like fungi, bacteria and viruses are a major challenge for crop improvement. The most important class of genes that has been used by breeders for disease control are the plant resistance (*R*) genes. The introgression of *R*-genes into elite cultivars *via* traditional breeding can take up to 15–20 years. This process can be considerably shortened by introducing these genes by genetic Modification of crop plants (Rommens and Kishore, 2000). Recently cloning of the major resistance gene *RB* in *Solanum bulbocastanum* by using a map-based approach in combination with a long-range (LR)-PCR strategy has been reported. A cluster of four resistance genes was found within the genetically mapped *RB* region. Transgenic plants containing a LR-PCR product of one of these four genes conferred broad spectrum late blight resistance to potato (Song et al., 2003).

Significant advances have also been made in the development of virus-resistant crop plants since the first report of pathogen-derived resistance (PDR) to limit virus infection in the form of coat protein mediated resistance (Beachy, 1997). Although coat protein-mediated resistance (CPMR) has been the most widely applied PDR strategy for the last 20 years, it could be replaced in future by the recently developed RNA-mediated mechanisms based on short interfering RNA (siRNA) and microRNA (miRNA)-mediated gene silencing- derived strategy (Ritzenthaler, 2005). Besides, Niu et al. (2006) reported that artificial miRNAs can confer virus resistance in plants.

2.3.5 Pharmaceuticals, Proteins, Plastics Etc

For more than two decades, bio-engineered plants have produced protein therapeutics for human and animal use. Almost all proteins produced by other existing systems, including antibodies, vaccines and plasma proteins, have now been manufactured in plants (Joshi and Lopez, 2005). Several seed crops, including cereals, grain legumes and oilseeds, have been explored as production platforms, and the first commercial products—all technical proteins and enzymes—have already reached the market (Stoger et al., 2005). Fibrous proteins from spiders, proteins with synthetic multiple repeats, mammalian structural proteins such as collagen and biodegradable plastic have been produced in transgenic plants (Scheller and Conrad, 2005). Environmental pollution with metals and xenobiotics is a global problem, and the development of phytoremediation technologies for the plant-based clean up of contaminated soils is therefore of significant interest (Krämer, 2005).

2.4. Safety of Transgenic Plants

Because of the widespread public concern regarding the safety of biotech crops, rigorous testing and safety assessment is carried out before any such crops are approved for commercial cultivation. Besides the concern for marker genes, other worries with genetically engineered crops persist partially because of the perception that gaining access to a wider range of genetic diversity, crossing species barriers, and introducing foods with additional proteins present safety concerns via our diets (Cohen et al., 2003). The likelihood that a transgene will spread in the environment depends on its potential fitness impact. Mathematical models and empirical experimental evidence suggest that genetic approaches have the potential to effectively prevent transgenes from escaping into wild relatives and becoming established in wild populations that are not reproductively isolated from genetically engineered crops (Lee and Natesan, 2006). In regards to food safety issues, most biotech products are considered substantially equivalent to traditional counterparts, with exceptions for certain well-defined differences. Safety evaluations usually focus on these defined differences (Cohen et al., 2003).

In the US, according to the National Academy of Sciences, the trait of a GM plant should be the focus of the risk assessment, and not the process by which it has been produced (NAS, 2000). As a result the Environmental Protection Agency (EPA), requires that risk assessment tests are performed on non-target organisms to assess toxicity of so called plant-incorporated protectants. This is in contrast to EU regulatory procedures which consider GM plants as something new, for which both direct (toxic) and indirect effects should be assessed prior to their commercial release. Since in the US the major concern regarding the potential effects of GM plants expressing insecticidal proteins on non-target organisms is that related to the toxicity of the protein contained in the plant, test requirements on non-target organisms remained similar to those required for plant protection products. This means testing the effects on a set of representative insects such as bees, ladybirds and lacewings (NAS, 2002; EPA, 2002) using a tiered testing system in which the first tier reflects a maximum hazard approach. Organisms are exposed to the pure insecticidal protein at a range of doses and concentrations 10 to 100 times higher than those expected in the environment. Since these tests do not provide any information on the possible non-intended indirect effects that could occur once the plants are released in the field, the EPA often requires information from field studies (EPA, 2003). Field studies, however, have often the drawback of being expensive and lack appropriate statistical power.

2.4.1. Effects of Transgenic Plants on Natural Ecosystems

Genetic pollution is undesirable gene flow into wild populations. The term is usually associated with the gene flow from a genetically engineered (GE) organism (or genetically modifed organism (GMO). to a non GE organism; however, conservation biologists and conservationists are using it to describe gene flow from a domestic, feral, non-native or invasive species to a wild indigenous population. Hybridization of crops with nearby weeds may enable weeds to acquire traits we wish they didn't have, such as resistance to herbicides. Research results indicate that crop traits may escape from cultivation and persist for many years in wild populations. Genes that provide a competitive edge, such as resistance to viral disease, could benefit weed populations around a crop field. Many cultivated crops have sexually compatible wild relatives with which they hybridize under favorable circumstances. The likelihood that transgenes will spread can be different for each crop in each area of the world. Thus each crop must be evaluated individually for the risk of gene flow in the area where it will be grown.

Many factors influence the potential for gene flow from crop to crop. Some crops are highly out crossing, with pollen carried to other fields by wind and by insects. Other species are highly self-pollinating, with little potential for pollen transfer to neighboring plants. Because of the differences among crops species, every case must be evaluated individually for potential to contribute to gene flow from transgenic to conventional crops.

2.4.2. Effects of Transgenic Plants on Human Health

Genetically modified (GM) plants of the ''first generation'' were conferred with traits beneficial to producers, whereas GM plants in subsequent ''generations'' are intended to provide beneficial traits for consumers. Golden Rice is a promising example of a GM plant in the second generation, and has overcome a number of obstacles for practical use.

Furthermore, consumer-acceptable plants with health-promoting properties that are genetically modified using native genes are being developed. The emerging technology of metabolomics will also support the commercial realization of GM plants by providing comprehensive analyzes of plant biochemical components.

Agronomic traits such as herbicide tolerance, insecticides, virus resistance, disease-tolerance, and delayed ripening, were the initial focus of genetic engineering in plants, which are primarily beneficial to growers and the seed companies that pioneered the work (Fig. 1). These are the so-called ''first generation genetically modified (GM) plants''. In 2005, almost all GM acreage was occupied by first generation plants, dominated by herbicide-tolerance (71%), followed by insect resistance (18%) and a combination of these traits (11%) (ISAAA). The last decade has largely been dedicated to the development of ''second generation'' GM plants with traits beneficial to consumers such as high nutritional value, premium quality and low allergenicity. The second generation includes soybeans with high oleic acid- or laurate-content and rice with high vitamin A or iron content, or low allergenicity. Development of a ''third generation'' as protein factories of vaccines and therapeutic medicines is in progress. First generation GM plants mainly benefited producers by improving crop productivity, reducing agrichemical inputs and subsequently reducing labor costs. However, GM plants provided no obvious benefits for consumers over traditional plants. In contrast, the next generation of GM plants has been developed based on consumer benefits. Researchers and developers have been carefully choosing the traits to induce, including their origin, expression systems and selectable markers. The second and third generation GM crops which can produce substances for promoting human health, including medicines and vaccines have great potential for commercial application (Keiko Yonekura-Sakakibara and Kazuki Saito, 2006).

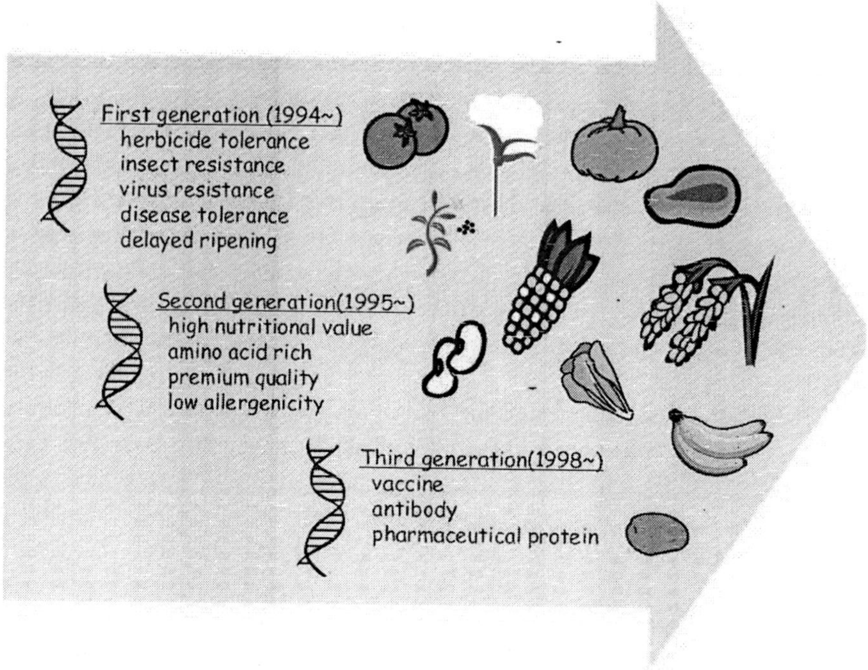

Figure 1. Generations of genetically modified plants. The traits of each generation are shown.

Fears and exaggerated claims of possible adverse effects of biotech foods and crops on humans and the environment are unfounded and have no basis in fact. Indeed, it is remarkable that no such ill effects have been documented after 12 years of extensive cultivation in diverse environments, and the consumption of biotech foods by more than a billion humans and even a larger number of animals. Many practical applications of plant biotechnology, including the introduction into the market place of other biotech crops with useful and highly desirable characteristics, have been prevented by political and ideological opposition from vested interests—including the European Union, a few developing countries, and some self-serving environmental groups—and by the many redundant, overly stringent, time-consuming and exorbitantly expensive regulatory requirements (Vasil, 2003).

The potential human health hazards from transgenic crops are also generally recognized by the scientific and regulatory communities (NRC, 2001). The potential risks generally relate to 'the possibility of introducing new allergens or toxins into food-plant varieties, the possibility of introducing new allergens into pollen, or the possibility that previously unknown protein combinations now being produced in food plants will have unforeseen secondary or pleiotropic effects.' Scientific studies have found transgenic crops that contain allergens (Nordlee et al., 1996) as well as crops with unintended effects. As with the potential environmental risks from transgenic crops, the likelihood that a particular transgenic crop may have a harmful health effect may be small but if it can occur, the government should step in to make sure that any transgenic crop with those harmful characteristics is not commercialized.

According to the information reported by the WHO, the genetically modified (GM) products that are currently on the international market have all passed risk assessments conducted by national authorities. These assessments have not indicated any risk to human health. For transgenic crops, protecting humans and the environment is the primary, but not exclusive, reason for government oversight and regulation. To date, there is little evidence of specific harmful environmental effects from the millions of acres of transgenic crops that have been planted worldwide nor any evidence of harm from the many foods that humans have eaten containing transgenic crop ingredients. There is general scientific consensus, however, that individual transgenic crops could potentially present risks to humans or the environment. For example, the United States National Research Council identified four categories of potential environmental hazards from the release of a transgenic crop: '(1) hazards associated with the movement of the transgene itself with subsequent expression in a different organism or species, (2) hazards associated directly or indirectly with the transgenic plant as a whole, (3) non-target hazards associated with the transgene product outside the plant, and (4) resistance evolution in the targeted pest population. That one or more of those risks could result from applying transgenic technology to a crop is generally recognized throughout the scientific community (Snow and Moran-Palma, 1997; NRC, 2001)

2.4.3. Effects of Biotechnology in Industries Product

The development of novel bioindustries is dependent upon the successful tapping of the invisible reservoir of microbial diversity. Naturally occurring or genetically engineered microorganisms are constituent components of bioactive processes that are conducive to the environment and the generation of market-competitive technologies. The role of microbial intervention in every sphere of human existence and endeavour is so integral that it is always glossed over. However, the role of microorganisms, and those that have been genetically engineered for specific purposes, in globalized biotechnologies is significant especially since they are economic agents of change, for example in fields such as: bioremediation bacterial ushing of aquatic and terrestrial environments; bioconversion technologies conversion of biowastes into utility products; biofuels and bioenergy generation biodiesel, alcohols, and ethers; biocleaning technologies desulphurization of coal; biopharmaceutical production gene therapy; bioproducts for agriculture biofertilizers, biopesticides; bioinformatics DNA-based computers, gene sequencing, novel plants and oral ornamentals.

2.4.4. Effects of Transgenic Biotechnology on Biodiversity

Fear for the loss of biodiversity is the focal point of opposition of several influential environmentalists groups against GM crops. It was hypothesised that GM crops could be a threat to the crop diversity or outgrow a local flora to the detriment of native species in Mexican maize landraces (Rissler and Mellon, 1993). However, a large-scale systematic survey showed no evidence for transgene introgression into maize landraces of Oaxaca, Mexico and is an important reference for local farmers, government agencies. The study also addresses socioeconomic and ethical implications of use of GM crops thereon (Garcia et al., 2005).We recognize that non-target herbivorous arthropods feeding on the transgenic crop may also be affected (lethally or sub-lethally). These effects may lead to a reduction in populations which may affect organisms in higher trophic levels. For registration of pesticide products, tests on non-target herbivorous arthropods are generally not required, despite the fact that most insecticides affect to some extent non-target herbivores that occur in the crop. By adopting the same principle, one can argue that risk assessment for non-target herbivores should not be required for registration of insect resistant transgenic plants. However, given the season-long expression of insecticidal proteins in transgenic plants it would be useful to have information on the effects of transgenic plants on some non-target herbivores.

One of the primary concerns related to the adoption of insect resistant transgenic plants in the environment is the detrimental effect that these may pose on non-target organisms, including entomophagous arthropods (parasitoids and predators) which have an important function in regulating pests. Despite the fact that regulatory bodies require information regarding the potential risk of releasing transgenic plants in the environment, to date, no specific protocols have been designed for assessing the risks of insect resistant transgenic crops on entomophagous arthropods. The only insect resistant transgenic plants that are commercially available are those expressing genes which code for *Bacillus thuringiensis* 612 ANNA DUTTON ETAL (Bt) toxins. The amount of area being cultivated with these crops is rapidly increasing (James, 2002), and other genes coding for new Bt-toxins, lectins, proteinase or α-amylase inhibitors, and other insecticidal products have been successfully engineered in plants. Some of these plants are being tested at the field scale, such as peas (*Pisumsativum*) expressing the gene coding for common bean α-amylase inhibitors (αAIs).

Future agricultural and, ultimately, also natural ecosystems will be challenged by the large-scale introduction of entirely novel genes and gene products in new combinations at high frequencies all of which will have unknown impacts on their associated complex of non-target organisms, i.e. all organisms that are not targeted by the insecticidal protein. In times of severe global decline of biodiversity, pro-active precaution is necessary and careful consideration of the likely expected effects of transgenic plants on biodiversity of plants and insects is mandatory. The future challenge in pest management will be to explore how transgenic plants can be incorporated as safe and effective components of IPM systems and what gene technology can contribute to the needs of a modern sustainable agriculture that avoids or reduces adverse impacts on biodiversity? For mainly economically motivated resistance management purposes, constitutive high expression of Bt-toxins in transgenic plants is promoted seeking to kill almost 100% of all susceptible target pest insects. However, for pest management this is usually not necessary. Control at or below an established economic injury level is sufficient for most pests and cropping systems. It is proposed that partially or moderately resistant plants expressing quantitative rather than single gene traits and affecting the target pest sub-lethally may provide a more meaningful contribution of agricultural biotechnology to modern sustainable agriculture. Some examples of such plants produced through conventional breeding are presented. Non-target effects may be less severe allowing for better incorporation of these plants into IPM or biological control programs using multiple control strategies, thereby, also reducing selection pressure for pest resistance development.

2.4.5. Un-Predictability of Transgenic Biotechnology

Transgenic biotechnology presents an exciting range of possibilities, from feeding the hungry to preventing and treating diseases; however, these promises are not without potential peril. Genetic engineering is changing the terrain of development studies. Societies have reacted with indifference or appreciation to genetically engineered pharmaceuticals, beginning with insulin; yet for food and agriculture, a globally contentious politics and unprecedented policy dilemmas have arisen. Transgenic organisms raise questions of property, ethics and safety unimaginable a generation ago: what can be owned and with what responsibility? Are we blurring the lines between species by creating transgenic combinations? What ethical, social, and legal controls or reviews should be placed on such research? Much turns on science: how one conceptualizes evidence, knowledge, uncertainty and risk. Both opponents and proponents of frontier applications in biotechnology have a poverty story to tell, but with divergent implications. The balance in this global debate has perceptibly shifted; a new developmentalist consensus concludes that the world's poor may benefit from genetic engineering: the question is 'under what conditions'? The new consensus assumes much about biosafety, bioproperty and biopolitics that is contrary to ground realities - the actual capacity of firms and states to monitor and control biotechnology - but raises new questions at the frontiers of development studies. Transgenic and genetic engineering present intriguing and difficult challenges for 21st century scientists and ethicists. Until we as a society or, perhaps, as a global entity can agree on what beings, human or otherwise, are worthy of moral and legal status and respect, we can expect intense cross-disciplinary debate and discussion as new intelligent life is created through science and medicine.

2.4.6. Safety Assessment and Management of Transgenic Plants

Transgenic crops have the potential to benefit both developed and developing countries. To ensure safe crops to humans and the environment, a strong, but not stifling, regulatory system needs to be established and properly implemented. Compositional analysis represents one component of the risk assessment process for GMOs (EFSA 2004). In Europe, the risk assessment strategy for GMOs focuses on comparison between the GMO and derived products and their non-GM counterparts. This comparative assessment approach is based on the premise that traditionally cultivated crops have gained a history of safe use as food/feed for human or animals and with regard to any adverse environmental impact. These crops serve as a baseline for the environmental and food/feed safety assessment of GMOs. The concepts of familiarity and substantial equivalence were developed by the OECD (EFSA 2004) and further elaborated by WHO/FAO. This comparison is the starting point of the safety assessment which then focuses on the environmental or food/feed safety and nutritional impact of any intended or unintended differences identified.

"Novelty" and "societal benefit" are two central criteria for the possible significance of technological developments. The present report on transgenic plants of the second and third generation (with the emphasis on molecular farming) is devoted to a balanced consideration of these two perspectives: In terms of novelty, the aim is to focus on new questions of evaluation, while the aspect of societal benefits will focus on the societal context as a whole, rather than one-sided consideration of the risk dimension or specific economic interests. At present, transgenic plants of the safety evaluation focused on two aspects, one is the safety of the environment, and the other is food security. There is little reliable information on the individual aspects of possible ecological and health risks of the various groups (or even individual transgenic properties). The risk discussion for FF GMP is focusing on the basic question of safety evaluation of innovative and primarily functional foods, while for PMP GMP the emphasis is on possible release into the environment and food and preventing this (depending on the individual instance, PMI can be closer to FF or PMP GMP). Particularly with regard to PMP, the risk debate on molecular farming generally has so far concentrated almost entirely on the question of reliable sequestration and containment/confinement of GMP. The discussion of benefits is taking on new priority compared with the first generation of GMP, including risk evaluation and regulation. So far, it has been possible to ignore doubts about the benefits of the genetically introduced properties e.g. from the regulatory point of view (because no concrete risks to health and the environment were established as a prerequisite for approval), and to leave evaluation to market forces. In future, however, the desired benefit (e.g. of drug production) is likely to play a greater role – at least in individual cases – in risk evaluation, including in the approval decision

To properly utilize agricultural biotechnology, a strong regulatory system is needed to ensure products are safe to humans and the environment. There is currently no country with an ideal regulatory system, although portions of many different country's systems are worth using as models. It is important that every country be given the opportunity to establish a biosafety system so that it can make its own determination whether a particular product meets its safety standards. Once a system is in place, two countries may rationally reach different conclusions on approving the same product, due to a different balancing of the risks and benefits of that product to that country's population and-environment. Each country needs a strong regulatory structure, however, to be able to make those decisions. It is essential that regulatory systems be transparent and allow for public participation. Unless the public is provided access to the system as well as detailed information about the products that go through it, there will be little public trust in the regulatory process and no consumer confidence in the marketed products. The key to reaping the long-term benefits of agricultural biotechnology is a strong, but not stifling, regulatory system that independently reviews and approves each product for human and environmental safety before it is commercialized.

3. Bioremediation in Environment Pollution

Bioremediation technique can be considered a promising alternative to reduce the toxicity of water, soil and petroleum pollutants. The term "bioremediation" has been used to describe the process of using microorganisms to degrade or remove hazardous components of the wastes from the environment (Glazer and Nikaido, 1995). Here, recent advances of bioremediation world water resources, soils and petroleum industry are discussed.

3.1. Bioremediation Technology and Bioreactors

The term "bioremediation" has been used to describe the process of using microorganisms to degrade or remove hazardous components of the wastes from the environment. A narrow definition of bioremediation might focus on the conversion of contaminating organic molecules to CO_2, water and inorganic ions, and the oxidation or reduction of contaminating inorganic ions. A broader definition would include biological processes for ameliorating extremes of pH, concentrating contaminants so that they can be more easily removed by physical techniques, converting toxic species to less toxic forms, thereby restoring functional ecosystems in contaminated or disturbed sites when the contaminants or disturbance cannot be removed (Adhya, 2008).

Bioremediation technology uses microorganisms to reduce, eliminate, contain and transform to benign products, contaminants present in soils, sediments, water or air. Micobial processes are beginning to be used in the cleanup of radioactive and metallic contaminants, though these contaminants present special problems since they cannot be destroyed, only transformed or contained. There is an increase in the types of contaminants to which bioremediation is being applied, including solvents, explosives, polycyclic aromatic hydrocarbons (PAHs), and polychlorinated biphenyls (PCBs). Bioremediation is an alternative to traditional remediation technologies such as landfilling or incineration. Bioremediation works by either transforming or degrading contaminants to non-hazardous or less hazardous chemicals. Bioremediation and biodegradation are of biofilm formation, bacterial adhesion and xenobiotic compounds. Environmental Biotechnology worked at the design phase and deals with mass transfer mechanisms in bacterial biofilms, surface and groundwater flow and contaminant transport, optimization of novel catalytic materials and absorbents, fluid mechanics and heat transfer of reacting single and multi-phase flows, forest micrometeorology and microclimatology, estuarine restoration and many other areas of application.

The fungus *Trichoderma harzianum produces homothallin* II, a nitrile with a broad activity against fungi and Gram-positive and Gram-negative bacteria (Faull et al.,1994), *Chromobacterium violaceum* ATCC 53434 produces the isonitrile antibiotic aerocyanidin that is primarily active against Gram-positive bacteria (Parker et al.,1988).A number of fungi and bacteria utilise the toxicity of cyanide compounds in a more overt way by producing and secreting antibiotic cyanide compounds to inhibit competitive organisms.

Polycyclic aromatic hydrocarbons (PAHs) are compounds of intense public concern due to their persistence in the environment and potentially deleterious effects on human, environmental and ecological health. Bioremediation, which exploits the metabolic potential of microbes for the clean-up of recalcitrant xenobiotic compounds, has come up as a promising alternative. Microorganisms have been found to degrade PAHs either via metabolism or cometabolism. In spite of their xenobiotic properties, a variety of genera of gram-positive and -negative bacteria, fungi and algae have been isolated and characterized for their ability to utilize PAHs. Different regulatory mechanisms have been identified in different PAH degrading bacteria. Analysis of the PAH catabolic genes is of interest in both fundamental and applied contexts. Approaches such as improvement in PAH solubilization and entry into the cell, pathway and enzyme engineering and control of enzyme expression etc. are in development. Several strategies could be employed to improve the induction of PAH catabolic operons under field conditions. One strategy could be the expression of the PAH degrading genes under the control of a constitutive promoter. For this the novel approach of promoter implantation by homologous recombination developed by Ohtsubo et al. (2003). Another strategy could be the identification of a cheap and non-toxic inducer, as demonstrated by Gilbert and Crowley (1997) for PCB degrading *Arthrobacter sp.* B1B. Genetic engineering of well characterized enzymes is a strategy for the creation of the enzymes with the desired properties. Biphenyls and PCBs could be exploited (a new naphthalene dioxygense from *Sphingomonas* CHY-1 which has the ability to attack PAHs composed of up to five rings. The use of such strains having broad substrate specificities is one strategy to improve the degradation of PAHs. Selection of an appropriate strain is vital for successful *in situ* bioremediation. Successful application of plant–microbe systems for rhizoremediation relies on *in situ* establishment of a high level of competence of the introduced bacteria. The inoculation of pollutant-degrading bacteria on plant seed can be an important additive to improve the efficiency of phytoremediation or bioaugmentation in the near future (Archana Chauhan et al., 2008).

Sarkar et al. (2006) has reported that the use of molecular biomarkers as diagnostic and prognostic tools for marine pollution monitoring. In order to assess the impact of highly persistent pollutants such as polychlorinated biphenyls (PCB), polychlorinated dibenzo–dioxins (PCDD), polychlorinated dibenzo–furans (PCDF), polynuclear aromatic hydrocarbons (PAH), tributyltin (TBT) and other toxic metals on the marine ecosystem a suite of biomarkers are being extensively used worldwide. Among the various types of biomarkers, the following have received special attention: cytochrome P4501A induction, DNA integrity, acetylcholinesterase activity and metallothionein induction. These biomarkers are being used to evaluate exposure of various species of sentinel marine organisms (e.g. mussels, clams, oysters, snails, fishes, etc.) to and the effect of various contaminants (organic xenobiotics and metals) using different molecular approaches biochemical assays, enzyme linked immuno-sorbent assays (ELISA), spectrophotometric, fluorometric measurement, differential pulsed polarography, liquid chromatography, atomic absorption spectrometry. The induction of the biotransformation enzyme, cytochrome P4501A in fishes (*Callionymus lyra, Limanda limanda, Serranus sp., Mullus barbatus*) and mussels (*Dreissena polymorpha*) by various xenobiotic contaminants such as PCBs, PAHs, PCDs is used as a biomarker of exposure to such organic pollutants. The induction of cytochrome P4501A is involved in chemical carcinogenesis through catalysis of the covalent bonding of organic contaminants to a DNA strand leading to formation of DNA adduct. Measurement of the induction of cytochrome P4501A in terms of EROD (7-ethoxy resorufin O-deethylase) activity is successfully used as a potential biomarker of exposure to xenobiotic contaminants in marine pollution monitoring. In order to assess the impact of neurotoxic compounds on marine environment the evaluation of acetylcholinesterase activity in marine organisms is used as a biomarker of exposure to neurotoxic agents such as organophosphorus, carbamate pesticides etc. Metallothioneins (MTs) are induced by toxic metals such as Cd, Hg, and Cu by chelation through cysteine residues and are used in both vertebrates and invertebrates as a biomarker of metal exposure. The measurement of the levels of DNA integrity in marine organisms such as Sea stars (Asterias rubens) from the North Sea and the marine snails (Planaxis sulcatus) from the Arabian Sea along the Goa coast exposed to environmental xenobiotic contaminants clearly indicated the extent and the nature of pollution at the sampling sites along coastal environment.

Bioreactor systems are becoming more important for the production of large quantities of proteins, particularly in cases where traditional sources are limited due to cost and/or availability. Bioreactors based on mammalian cell lines are currently the system of choice for the production of proteins for therapeutic use and 21 different diagnostic and therapeutic proteins have been approved in the USA (Kelley, 2001). Plant-based bioreactors to be commercially viable, plants are not hosts to major human pathogens that can contaminate bacterial and mammalian systems (Herbers and Sonnewald 1999). PMP which was for the vaccine against ''Newcastle Disease Virus'' was a landmark success in the manufacturing of plant cell culture based proteins. To date, about 20-plant made proteins (PMPs) have been produced in proof-of principle studies. For plant cell and tissue cultures, a variety of bioreactor types providing growth and expression of bioactive substances is available today. In contrast to traditionally grown ''whole wild plants'' or ''whole transgenic plants'', their production in bioreactors guarantees defined controlled process conditions and therefore minimizes or even prevents variations in product yield and quality, which simplifies process validation and product registration. Most of the plant cell suspension cultures and hairy roots used are grown in submerged stirred bioreactors, bubble columns, airlift reactors and their modifications, gas-phase bioreactors and their combinations, or socalled hybrid systems up to m^3-range. The development of a number of novel bioreactor systems based on a diverse variety of living cells and organisms have been seen recently. Choice of bioreactor system for the production of a particular protein will help not only to enhance and sustain high productivity, but also to reduce process costs (Regine Eibl and Dieter Eibl, 2008).

3.2. In Situ Bioremediation

Bioremediation is a biodegradation process in which sites contaminated with xenobiotics are cleaned up by means of microbial bio-geochemical processes.

In general, remediation techniques can be divided into two classes depending on whether the pollutant is directly removed/degraded in-place or not, i.e., in situ or ex situ. One of the main problems associated with ex situ remediation is its high operating cost for activities like soil excavation and groundwater pumping. Ex situ technologies for the bioremediation of soil contaminated with toxic organic compounds include land farming, composting and bioreactors. Operation of these bioremediation strategies involves the excavation and or agitation of soil, which often leads to enhanced abiotic loss of contaminants. In recent years, in situ techniques have become popular. In situ treatments, including natural attenuation – defined as *in situ* attenuation processes that reduce the hazard associated with the release of contaminant (Soo Cho et al.,1997), biostimulation and bioaugmentation, usually avoid excavation and emission control costs, and may therefore be considered as cheap and clean technologies. In situ biodegradation has been shown to be effective at degrading or transforming a large number of organic compounds to environmentally acceptable or less mobile compounds. Soluble organic contaminants are particularly amenable to biodegradation. Relatively insoluble contaminants may also be degraded if they are accessible to microbial degraders (Chena et al., 2003) In situ and on site bioremediation of contaminated soils and groundwater has recently received widespread attention. A number of structure-biodegradability relationships have been developed for aerobic microbial metabolism which aid the assessing the potential of specific contaminants for bioremediation

Potential evidence for in-situ biodegradation is often gathered by comparing geochemical and microbial characteristics in the contaminated zone to those of uncontaminated background areas. Some types of evidence for in- situ biodegradation includes the following observations: (i) decreasing contaminant concentration over time, (ii) consumption of electron acceptors and production of their reduced byproducts, (iii) production of signature degradation intermediates and products, (iv) stable isotopic ratios and radioisotopic content of contaminants and degradation products, and (v) distinct patterns of elevated biomass concentrations and activities. While these observations can provide valuable information supportive of in-situ biological activities, each has inherent interpretational limitations, and, when used alone, can yield ambigious results. However, when used in combination, these observations can provide convincing evidence for the occurrence of in-situ bioremediation .in-situ bioremediation can be an effective method for treating fuel and other hydrocarbon contamination in soil and groundwater (Vishvesh K, 2002).

Sanjeet (2001) validates the large-scale use of a carrier-based bacterial consortium and nutrients for the treatment of land contaminated with oily sludge, a hazardous hydrocarbon waste generated by petroleum industry. Improper disposal and handling of oily sludge contaminates soil and may pose a serious threat to groundwater, and in situ bioremediation offers a promising means to reclaim such contaminated soil .The process treats the contaminants on site (in situ) instead of merely transporting them elsewhere.

3.3. Bioremediation of World Water Resources

Water plays without a doubt critical role in human health, agricultural production and modulating climate. Water also provides planetary biogeochemical processes with a medium for the distribution and mineralization of pollutants. Risks of water contamination from current and past pollutant discharges abound, for example, it is estimated that the world industrial output of organic chemicals to freshwaters alone totals over 19,000 metric tons per day. Population growth and economic development without industrial efficiency improvements and sound sewage treatment and management lead to increased pollutant loads to surface and groundwater. Water is a sustainable resource if properly managed. Several strategies are currently being developed on biological wastewater treatment.

There is a growing interest in the use of aquatic or semi-aquatic plants for the removal of heavy metals from contaminated aqueous streams. Such plants could be utilized for the improvement of water quality and for reducing the pollution load in water bodies. There has been reported that water hyacinths (*Eichornia crassipes*), pennywort *(Hydrocotyle umbellate* L.) (Dierberg et al., 1987), duckweed (*Lemna minor*) and water velvet (*Azolla pinnata*) (Jain et al., 1989) take up Pb, Cu, Cd, Fe and Hg from contaminated solutions. Kara (2004) analyzed bioaccumulation of copper from contaminated wastewater by using *Lemna minor* under laboratory conditions. A substantial decrease of the exposure in the somes River could be observed due to the upgrade of the municipal wastewater treatment plant (WWTP) in Cluj-Napoca. After the upgrade of the WWTP Cluj-Napoca, the concentrations in the somes of caffeine, pentoxifylline, cyclophosphamide, galaxolide, and tonalide were significantly reduced (over 75%) (Zaharie et al., 2008).Large-scale plants usually apply coagulation with aluminium or iron salts followed by filtration but a number of anions (e.g. nitrate) have very little tendency to coordinate with metal ions and low potential for co-precipitation (Duan and Gregory, 2003).

The use of membranes is particularly attractive for separating ions between two liquid phases (purified and concentrated water streams). Because many of the difficulties associated with precipitation, coagulation or adsorption and phase separation can be avoided. When membrane(s) are integrated with an appropriate bioprocess in a single process, the configuration is generally referred to as ''membrane bioreactor''. Membrane processes and membrane bioreactors include Reverse osmosis (RO), Nanofiltration (NF), Pressure-driven membrane bioreactors and, Dialytic membrane bio/processes, Electrodialytic membrane processes and Gas-transfer membrane bioreactors. Membrane technologies are already successfully used on large-scale for removal of inorganic anions such as nitrate, fluoride, arsenic species, etc. Membrane bioreactors allow for complete depollution but water quality, insufficiently stable process operation, and economical reasons still limit their wider application in drinking water treatment. The nitrate removal efficiency was close to 98% (Svetlozar et al., 2004). RO and NF can be used for removal of inorganic anions as single processes. It was demonstrated that Reverse osmosis (RO) could be effectively applied for removal of nitrate along with water desalination in a rural area (Schoeman and Steyn, 2003).

Anaerobic processes have been extensively applied for the digestion of primary and secondary sludge in wastewater treatment plants. Aanaerobic processes have played an important role in the organic load abatement and sludge manageability, mainly in large conventional aerobic treatment plants. From 1950 to 1980, several anaerobic reactor configurations were developed for industrial wastewater treatment (McCarty, 1982). Domestic anaerobic technology acts as a tool to promote a sustainable sewage treatment (DST). The ideal situation for DST systems would be the complete removal of pathogens (health protection) and the highest removal of COD (environmental protection) with recovery of energy (methane or hydrogen) and compounds of interest: nitrogen (as NH_4^+, NO_2), and NO_3), phosphorus (as phosphate) and sulfur (as S^0). The increasing use of anaerobic reactors as the first unit of DST systems is mainly due to the successful use of UASB reactors for industrial wastewater treatment. The earlier reports on the application of UASB reactors for DST are from the beginning of the eighties (Lettinga et al., 1982). At the moment, the UASB reactor is undoubtedly the most successful reactor for the treatment of raw domestic sewage, due to both the absence of a fixed bed (avoiding clogging) and the presence of a SLG separator on the reactor's top that prevents excessive solid losses. The recovery of nitrogen is quite complex due to the high solubility of all nitrogen forms of interest (NH_4^+, NO_2), NO_3). Scientists and practitioners pursue ways to remove nitrogen in its dissolved form from anaerobic effluents. Adsorption seems to be the most appropriated method for nitrogen recovery adsorption of nitrogen in its more reduced form (NH_4^+) is advantageous because this form predominates in anaerobic reactor effluents. For nitrogen removal from anaerobic effluents of DST in the traditional nitrification/denitrification design, there is a need for an external electron donor. This would open the opportunity to explore the use of electron donors produced in anaerobic reactors. Among such electron donors, VFA, methane, ammonia and sulfide are the natural candidates. Enhanced biological phosphorus removal (EBPR) has been achieved under alternating anaerobicaerobic conditions using polyphosphate-accumulating organisms (PAO) in the presence of an easily assimilating organic carbon source (Comeau et al., 1986; Eugenio et al., 2006).Based on conventional aerobic systems. The characterization and quantification of microorganisms in bioreactors is essential for describing and optimizing the bioconversion process. Most of the effort is done on the removal of more specific contaminants such as nutrients (N and P) and sulfurous compounds since they are becoming of great concern due to its impact on the quality of water bodies. There have been two recent discoveries of microbial conversions of nitrogenous compounds. One depends on the capability of ammonia oxidizers of denitrify under certain conditions resulting in a new one-step method for the removal of N-compounds. The second has been named the ANAMMOX process, wherein ammonium is oxidized to dinitrogen gas with nitrite as the electron acceptor; Biological phosphorous removal are based on the capacity of some denitrifying microorganisms to store ortho-phosphate intracellular as poly-phosphate in the presence of nitrate; The microbial conversions of sulfurous compounds involve the metabolism of several different specific groups of bacteria such as sulfate reducing bacteria, sulfur and sulfide oxidizing bacteria, and phototrophic sulfur bacteria (Santiago et al., 2004). *Nitrosomonas eutropha* was able to grow under anoxic conditions using hydrogen or organic compounds (acetate) as electron donor and nitrite as electron acceptor. The anaerobic biological treatment of industrial wastewater highly loaded with organic matter, nitrogenous and sulfurous compounds may induce manifold microbial scenarios. In consequence, very complex microbial consortia may grow within anaerobic reactors treating this type of complex matrixes of substrates (Lettinga, 1995). The use of anaerobic processes as the core technologycan improves the sustainability of DST (Eugenio et al., 2006).

Cyanides are a group of compounds that present in the environment in several forms including; hydrogen cyanide; simple inorganic salts including NaCN and KCN; complex metal cyanides, thiocyanates, and as nitriles where the CN binds to an organic radical (Kjeldsen 1999). According to the degree of cyanide liberation by these compounds, the toxicity of metal-cyanide complexes or nitriles vary .Free cyanide (HCN or CN)) is an extremely potent metabolic poison in toxicity, (Figueira et al. 1995). As a consequence of its widespread use in industry, and because many cash and food crops naturally synthesise cyanide compounds, there are large volumes of waste generated during industrial and agricultural production and processing that contain significant concentrations of cyanide. Associated with the discharge of municipal wastewater by sewage treatment plants, particularly from facilities that serve industrialised areas, where cyanide can reach 0.1 mg l^{-1} (Lordi et al., 1980).These must be treated before being discharged to prevent environment degradation. There are several widely used physical and chemical treatments of cyanide wastes that have achieved acceptance in commercial applications (Ebbs, 2004). A number of fungi and bacteria utilise the toxicity of cyanide compounds in a more overt way by producing and secreting antibiotic cyanide compounds to inhibit competitive organisms. The biological treatment of metal-cyanide contaminated wastewaters has been successfully demonstrated in laboratory scale experiments under aerobic and anaerobic conditions.

Relatively simple batch reactor set ups have exploited fungal biomass to degrade model metal cyanide containing wastewaters (Ezzi and Lynch, 2005).The biocatalyst 'Cyanidase', developed by Novo Nordisk, utilises an immobilised preparation of Alcaligenes denitrificans and has been developed for optimum cyanide degradation and stability with respect to the pH, ionic strength and presence of metal ions in cyanide contaminated wastewaters (Basheer et al., 1992). Microorganisms have been shown to possess a variety of catabolic activities that can be harnessed to remediate inorganic cyanide and nitriles. Further work is required to optimise the use of such biological systems both in wastewater .In particular, the development of microbial processes that are effective and robust in the face of extreme environmental conditions, such as low pH and co-contaminant toxicity, is required to ensure a technology that is competitive with the current chemical and physical remediation strategies currently practised to combat cyanide pollution (Joanne Baxter and Stephen Cummings, 2006).

A considerable amount of agricultural waste such as coconut husk is a highly lignocellulosic material. Parimala et al. (2004) have reported the removal of chromium from coastal water using five different products prepared from coconut husk. Nitrogenous contaminants such as ammonia and nitrite have been found to appear frequently in the source water used for coastal aquaculture (Krishnani et al., 2003). Parimala et al. (2004) has been carried out to investigate the use of coconut husk in the remediation of ammonia and nitrite from coastal water, and to examine their effect on other water quality parameters. Their result shows that coconut husk can be used for the effective removal of ammonia and nitrite from coastal water. The removal of ammonia was found to depend on materials, dose and time. The effect of five different types of materials from coconut husk at 1, 3 and 6 g/L on the removal of 1.203 mg/L ammonia presented that ammonia removal was most effective with raw coconut husk fiber (CRF) followed by oven dried coconut husk powder (COP) and charred coconut husk powder; The effect of five different types materials prepared from coconut husk at 1, 3 and 6 g/L on the removal of 1.024 mg/L nitrite shows that ammonia removal was effective with raw coconut husk fiber (CRF) followed by powder charred with acid (CSP). Oven-dried coconut husk powder (COP) and charred coconut husk powder (CCP) were also found to be effective. Other water quality parameters, such as salinity, did not show much change in the presence of CRF, CDF, COP, CCP and CSP throughout the course of the experiment.

Bioaccumulation of selenium (Se) in wetlands and evaporation ponds has created serious hazards to fish and waterfowl. Agricultural drainage water is the main source for Se contamination in this region. Treatment technologies currently being evaluated to reduce the Se load in agricultural drainage water include physical, chemical and biological methods. The concept of the ABSR process is to grow microalgae in drainage water at the expense of nitrate and then to utilize the naturally settled algal biomass as a carbon source for native bacteria which, in the absence of oxygen, reduce the remaining nitrate to nitrogen gas and reduce selenate to Se. The insoluble Se is then removed from the water by sedimentation in deep ponds and, as needed, by dissolved air flotation and sand filtration. Supplemental carbon sources such as molasses can be employed as a reductant in addition to algal biomass. Several species of microalgae isolated from agricultural evaporation ponds transform Se oxyanions into volatile alkylselenides, selenonium ions, and proteinaceous selenomethionine (Fan et al., 1997). In particular, two cyanophyte strains isolated from the evaporation basins of the Tulare Lake Drainage District (TLDD), California were very active in Se volatilization. Up to 70% of the Se was depleted from the medium containing 20–1,000 μg/L Se by the algal activity and a major fraction of the Se loss was accounted for via the volatilization route. It is possible that the algal Se volatilization activity may be related to the recent observation that waterborne Se concentrations at one of the TLDD basins were not increasing with rising salinity and that this trend was consistent year-around. Additional research is needed to determine this relationship. Several pilot-scale projects that use anaerobic reduction reactions to precipitate Se from drainage water have been tested. These treatment processes vary based on reactor design and the type of carbon source and nutrients provided to the microorganisms. Reactor designs include ground tanks containing sludge beds, fluidized beds, or fixed films. The advantages of the biological precipitation reaction are: soluble Se can be reduced by up to 95%; low production of solid residuals for disposal due to slow anaerobic growth rates; low required detention time allows use of standard engineering sizes of treatment systems, and the system is enclosed with no exposure to the environment or wildlife. The major disadvantage of this technology is that soluble Se residual of 30 ppb or greater is likely from biological treatment only and organic carbon costs are high unless low cost alternatives can be used. There is evidence that significant losses of Se, through biologically mediated methylation and volatilization, can occur in wetlands. Constructed wetlands constitute a complex ecosystem, the biological and physical components of which interact to provide a mechanical and biogeochemical filter capable of removing many different types of contaminants from water. Constructed wetlands remove Se by reduction to insoluble forms which are deposited in the sediments, by accumulation into plant tissues, and/or volatized to the atmosphere, i.e., through biological volatilization of plants, plant/microbe associations and microbes alone. Allen (1991) tested the ability of a constructed wetland to remove Se from water spiked with selenite and found that the rate of Se removal by the wetland exceeded 90% of the total Se introduced via biological volatilization as an important component of this loss. Actual field measurements of Se volatilization performed by Terry and his co-workers have shown that biological volatilization may have accounted for as much as 10–30% of the total Se removed from the San Francisco Bay constructed wetland (Hansen et al., 1996). Constructed subsurface flow (SSF) treatment wetlands can also be used to remove Se from irrigation drainage water. The wetland would consist of gravel beds that serve as the rooting medium for salt-tolerant emergent wetland plants such as bulrush, with the agricultural drainage water flowing through the gravel. As the water flows through the gravel bed, oxidation-reduction processes in the plant root zone (especially due to the interactions of the microbial community and the plant roots) would remove Se through a combination of uptake and volatilization. The primary advantage of SSF treatment wetlands is that they reduce the potential for an aquatic food-chain pathway for Se exposure to fish or birds.

There are a number of reports that one, more or algal assemblage could be used as organism indicative of water quality (Mohapatra and Mohanty, 1992). Use of such organism offers a great promise in ensuring their success as qualitative and quantitative indicator suitable for a wide variety of pollutants. Rai et al. (2005) have shown that different groups of algae growing on tannery effluent exhibited high Cr accumulation and have been found useful for bioremediation of Cr pollution Algae constitute important components in polluted and unpolluted fresh water bodies and are sensitive to metallic pollutants (Whitton et al. 1982), the adverse impact of any pollutant can be adjudged by their presence/absence and metal accumulation potential. Rai et al. (2008) demonstrates a relationship between physicochemical characteristics of these water bodies and metal bioaccumulation in major algal forms. The water bodies of Lucknow, Unnao and Kanpur (U.P.), India polluted through various point and non point sources were found to be either eutrophic or oligotrophic in nature. Maximumamount of Fe was found accumulated by species of Oedogonium sp. II (20,523.00 μg g^{-1} dw) and Spirogyra sp. I (4,520.00 μg g^{-1} dw), while maximum Chromium (Cr) was found accumulated in Phormedium bohneri (2,109.00 μg g^{-1} dw) followed by Oscillatoria nigra (1,957.88 μg g^{-1} dw) and Oedogonium sp. I (156.00 μg g^{-1} dw) and Ni in Ulothrix sp. (495.00 μg g^{-1} dw). Results showed that some of these forms growing in polluted environment and accumulating high amounts of toxic metals may be used as bioindicator species, their performance in metal contaminated water under different ecological niche is to be ascertained.

3.4. Bioremediation of Soils

Bioremediation is considered to be a safe, efficient, eco-friendly and economic means of removing pollutants from contaminated soil.

Endosulfan is used extensively throughout the world to control the insect pests of a wide range of crops. It can bind to soil particles and persist for a relatively long period with half-life of 60-800 days (Siddique et al., 2003). Endosulfan is extremely toxic to fish and aquatic invertebrates and has been implicated increasingly in mammalian gonadal toxicity, genotoxicity, and neurotoxicity (Sutherland, 2000). Several microorganisms both bacteria and fungi, which degrade endosulfan have been isolated and characterized report on the diversity of endosulfan degrading bacterial systems, evaluate the capability of the enriched bacterial isolates to *in situ* degrade endosulfan and endosulfan sulfate in the complex environment of soil and to assess the survival of the introduced population in the soil based on ERIC - PCR. Molecular characterization based on 16s rDNA gene sequence analysis of bacterial colonies isolated from endosulfan contaminated soil showed the presence of *Ochrobacterum* sp, *Burkholderia* sp, *Pseudomonas alcaligenes, Pseudomonas* sp and *Arthrobacter* sp which degraded 57-90% of α-endosulfan and 74-94% of β-endosulfan after 7days (Mohit et al., 2008).

Earthworms maintain aerobic conditions through the continuous mixing of the soil (Schack-Kirchner and Hildebrand, 1998). They also ingest soil and expel a partially stabilized product (casting), in this way they ensure the availability of organic substrates for proliferation of the autochthonous microorganisms in the soil, thus increasing the microbiological and biochemical soil activity (Wolters, 2000).The use of earthworms, as an efficient method to support the bioremediation of a soil has been experimented (Singer et al., 2001). The bioremediation treatments were effective in the degradation of hydrocarbon polluted compounds. In particular, the use of just mineral nutrients to stimulate soil microbial activity caused a high decrease in total organic carbon, thus suggesting an attack of the native soil organic matter (priming effect). On the other hand, the addition of compost (especially with earthworms) had the double effect of adding nutrients and labile organic matter for the soil authoctonous microrganisms, but also of introducing new microrganisms in the soil system. Finally, the compost treatments, stimulating soil metabolic processes, achieved the best result in the hydrocarbon degradation preserving the biochemical quality of the soil (Brunello, 2006).

Katia et al. (2005) reported the ability of tropical Brazilian basidiomycetes to degrade pentachlorophenol (PCP) in soils from areas contaminated with organochlorine industrial residues. Thirty-six basidiomycetes isolated from different tropical ecosystems were tested for tolerance to high PCP concentrations in soil. *Peniophora cinerea* and *Psilocybe castanella,* two strains of *Trametes villosa*, *Agrocybe perfecta*, *Trichaptum bisogenum* and *Lentinus villosus* were able to colonize soil columns containing up to 4600 mg pentachlorophenol kg^{-} soil. The study shows the potential of Brazilian fungi for the biodegradation of toxic and persistent pollutants and it is the first to report fungal growth and PCP depletion in soils with high pentachlorophenol concentrations.

Polycyclic aromatic hydrocarbons (PAHs) represent a large and diverse group of organic molecules with low bioavailability, hydrophobic nature, strong sorption phenomena, and high persistence in soil. PAH-contaminated soil employing various remediation technologies such as solid phase treatment, land treatment/farming, composting, bioreactors, phytoremediation, enzyme catalyzed bioremediation and combined methods (chemical pre-treatment followed by bioremediation) has been documented, in addition to the application of strategies such as biostimulation, microbial adaptation, bioaugmentation, bacterial chemotaxis, etc. (Venkata et al., 2006).

Phytoremediation, or rhizoremediation, is the use of plants and/or the associated rhizosphere to decontaminate polluted sites, and is considered today to be a realistic, low-cost alternative for treating extensive areas of pollution by organic chemicals (Susarla et al., 2002). The advantages of phytoremediation are that it preserves the natural structure and texture of the soil (Huang et al., 2004). Two reactor configurations, namely soil column bioreactors and a soil slurry phase bioreactor were studied in the application of bioreactor technology for the remediation of PAH-contaminated soil (Kumada et al., 2005).

Metal contamination of soil is especially problematic because of the strong adsorption of many metals to the surfaces of soil particles. Metals and radionuclides cannot be biodegraded. Bioremediation of metals and radionuclides relies on a complex interplay of biological, chemical and physical processes. There are at least three types of microbial processes that can influence toxicity and transport of metals and radionuclides: biotransformation, biosorption and bioaccumulation (Henry et al., 2005). Bioremediation of metals and radionuclides has had many field tests, Field research in this area has occurred for many different metals and radionuclides using different strategies characterized in six major categories: biotransformation, bioaccumulation/bisorption, biodegradation of chelators, volatilization, treatment trains, and natural attenuation (Terry et al., 2005). The biotransformation of metals is an exciting, developing strategy to treat metal contamination, especially in environments that are not accessible to other remediation technologies. Remediation of contaminated sediments and soil requires detailed in situ characterization of the speciation of the toxic substances and their transformations with respect to time and spatial distribution (Eric et al., 2005).

Siderophores are small molecular weight extracellular organic compounds secreted by microorganisms under iron-starved conditions, used by them to chelate and solubilize iron. Though they are specific ferric iron chelator, but is reported that they bind other metals also, such as divalent heavy metals and actinides because of potentially high metal-siderophore stability constants. Thus metal contaminant fate and transport in subsurface environment can be heavily influenced by siderophores. This approach can be successfully used in removing many toxic metals off the soil which poses a serious health threat. Anupa et al.(2007) reported that the siderophores produced by P. azotoformans was detected by CAS assay and determination of its functional groups has revealed that it is of mixed type i.e., catecholate as well as hydroxamate type; Siderophore showed high affinity for complexation with several metals in the following order —Cd^{2+}=Ni^{2+}=Pb^{2+}=Sn^{2+}=AsO_4^{-2}=AsO_2^{-1}>Mn^{2+}=Co^{2+}=Cu^{2+}=Fe^3 and very weakly binds Al and Cr.3. The batch studies reported herein provides encouraging and promising results and presents an ecofriendly approach for removing the arsenic by siderophore from contaminated soils. Five washings by siderophore, EDTA, CA removed almost 92.8%, 77.3%, 70.0% arsenic respectively as compared to only 33.8% removal by control. Washing of arsenic contaminated soil with tap water revealed that ≈65.8% of arsenic in contaminated soil is in freely available or weakly bound form. The use of siderophore for decontamination of arsenic off the soil showed no toxic effect on soil microbial population. These results confirm the potential applicability of siderophore for metal bioremediation. The contaminated soil has arsenic above the legislative limit and the treatment with siderophore could bring it down to a lower level, but still complete removal needs extended washings with the siderophore. Bonding structure studies by FTIR spectroscopy revealed hydrogen bonding between soil, arsenic and siderophore.

George et al. (2006) propose the use of solid polymer beads for the removal of a semi-volatile model contaminant (phenol) from soil and its subsequent destruction in a bioreactor via a microbial consortium isolated for its ability to degrade phenol. Polymer beads have been used to absorb high concentrations of phenol from soil decreasing the initial concentration of 2.3 g kg^{-1} soil to 100 mg kg^{-1} soil and achieving a phenol loading within the polymer beads of 27.5 mg phenol g^{-1} beads. The phenol-loaded polymer beads were removed from the soil and placed in a bioreactor, which was then inoculated with a phenol-degrading microbial consortium. All of the phenol contained within the polymer beads was shown to desorb from the polymer matrix and was degraded by the microbial consortium. The beads were used again (twice) in a similar manner with no loss in performance. Polymer beads have been demonstrated to readily absorb high levels of phenol from soil reducing the concentration to levels that may result in activity of indigenous organisms. The loaded polymer beads, used in conjunction with a solid liquid partitioning bioreactor, were shown to be an effective method for the ex situ bioremediation of phenol contaminated soil.

Bioremediation of pesticide in soil was documented by Dileep (2008). Microbes were used to degrade the pesticides in situ. A successful bioremediation technique requires an efficient bacterial strain that can degrade largest pollutant to minimum level. The rate of biodegradation in soil depends on four variables i.e. availability of pesticide or metabolite to the microorganisms; physiological status of the microorganisms; survival and/or proliferation of pesticide degrading microorganisms at contaminated site and sustainable population of these microorganisms. Biodegradation is a natural process, where the degradation of a xenobiotic chemical or pesticide by an organism is primarily a strategy for their own survival. Most of these microbes work in natural environment but some modifications can be brought about to encourage the organisms to degrade the pesticide at a faster rate in a limited time frame. This capability of microbe is some times utilized as technology for removal of contaminant from actual site. Knowledge of physiology, biochemistry and genetics of the desired microbe may further enhance the microbial process to achieve bioremediation with precision and with limited or no scope for uncertainty and variability in microbe functioning. Several bacteria competent of degrading either carbofuran, carbarly, baygon or aldicarb have been isolated and characterized. The degradation of HCH and distribution of *lin* gene in

Sphingomonads. *S. indicum* B90A was found to contain two non-identical *lin*A genes(Lal et al., 2006).The gene Ese encoding enzyme from monooxygenase family capable of degrading both endosulfan α and β from *Arthrobacter* sp (Weir et al., 2006). Esd gene that had same sequence homology to monooygenase family that use reduced fl avin, provided by a separate fl avin reductase enzyme, as cosubstrates in *Mycobacterium smegmatis* (Sutherland et al., 2002).Gene encoding for enzyme has been identified for several pesticides, which will provide a new inputs in understanding the microbial capability to degrade a pesticide and develop a super strain to achieve the desired result of bioremediation in a short time. Endosulfan(1,2,3,4,7,7-hexachlorobicyclo-2,2,1-heptene–2,3-bishydroxy methane -5,6-sulfite) is a mixture of two stereoisomers, α- and s-endosulfan, in a ratio of 7:3. Endosulfan is used extensively throughout the world to control the insect pests of a wide range of crops and most importantly, it has been used for the control of *Helicoverpa* sp. in cotton cropping. Endosulfan is extremely toxic to fi sh and aquatic invertebrates and has been implicated increasingly in mammalian gonadal toxicity, genotoxicity, and neurotoxicity (Sutherland et al., 2000). Molecular characterization based on 16s rDNA gene sequence analysis of bacterial colonies isolated from endosulfan contaminated soil showed the presence of *Ochrobacterum* sp, *Burkholderia* sp, *Pseudomonas alcaligenes, Pseudomonas* sp and *Arthrobacter* sp which degraded 57-90% of α-endosulfan and 74-94% of β-endosulfan after 7days. Whole cells of *Pseudomonas sp* and *Pseudomonas alcaligenes* showed 94 and 89% uptake of α-isomer and 86 and 89% of β- endosulfan respectively in 120 min. In *Pseudomonas* sp, endosulfan sulfate was the major metabolite detected during the degradation of α-isomer, with minor amount of endosulfan diol while in *Pseudomonas alcaligenes* endosulfan diol was the only product during α-endosulfan degradation. Whole cells of *Pseudomonas sp* also utilized 83% of endosulfan sulfate in 120 min. *In situ* applications of the defi ned consortium consisting of *Pseudomonas alcaligenes* and *Pseudomonas* sp (1:1) in plots contaminated with endosulfan showed that 80% of α-endosulfan and 65% of β-endosulfan was degraded after 12 weeks of incubation. Endosulfan sulfate formed during endosulfan degradation was subsequently degraded to unknown metabolites. ERIC-PCR analysis indicated 80% survival of introduced population of *Pseudomonas alcaligenes* and *Pseudomonas* sp in treated plots (Mohit Kumar et al., 2008).

The capacity of basidiomycetes to degrade Pentachlorophenol (PCP), as well the possibility of using these organisms in processes of reclamation of PCP-contaminated soils, have been the subject of much previous research. Thirty-six basidiomycetes isolated from different tropical ecosystems were tested for tolerance to high PCP concentrations in soil. *Peniophora cinerea* and *Psilocybe castanella*, two strains of *Trametes villosa, Agrocybe perfecta, Trichaptum bisogenum* and *Lentinus villosus* were able to colonize soil columns containing up to 4600 mg pentachlorophenol kg^{-1} soil. The first four species were inoculated into soil containing 1278 mg pentachlorophenol kg^{-1} soil supplemented with gypsum and sugar cane bagasse. P. cinerea, *P. castanella, T. villosa* CCB176 and CCB213 and *Agrocybe perfecta* reduced the PCP present in the contaminated soil by 78, 64, 58, 36 and 43%, respectively, after 90 days of incubation. All fungi mineralized [^{14}C] pentachlorophenol, mainly *P. cinerea* and *T. villosa* with the production of 7.11 and 8.15% $^{14}CO2$, respectively, during 120 days of incubation. All fungi produced chloride ions during growth on soil containing PCP, indicating dehalogenation of the molecule. Conversion of PCP to pentachloroanisole was observed only after 90 days of incubation in soils inoculated with *A. perfecta, P. cinerea* and one of *T. villosa* strain. The present study shows the potential of

Brazilian fungi for the biodegradation of toxic and persistent pollutants and it is the first to report fungal growth and PCP depletion in soils with high pentachlorophenol concentrations(Ka′ tia Maria et al.,2005).

The development of effective in situ and on-site bioremediation technologies can facilitate the cleanup of chemically contaminated soil sites. Knowledge of biodegradation kinetics and the bioavailability of organic pollutants can facilitate decisions on the efficacy of in situ and on-site bioremediation of contaminated soils and determine the attainable treatment end-points. Two kinds of compounds have been studied: phenol and alkyl phenols, which represent hydrophilic compounds, exhibiting high water solubility and moderate to low soil partitioning; and polycyclic aromatic hydrocarbons which are hydrophobic compounds with low water solubility and exhibit significant partitioning in soil organic carbon. Representative data are given for phenol and naphthalene. The results provide support for a systematic multi-level protocol using soil slurry, wafer and porous tube or column reactors to determine the biokinetic parameters for toxic organic pollutants. Insights into bioremediation rates of soil contaminants in compact soil systems can be attained using the protocol. A three-step experimental protocol for determining important kinetics parameters for the *in situ* biodegradation of toxic chemicals in soils was developed using phenol and naphthalene as the test compounds. The protocol was developed so that the experimental schemes used grew in complexity toward the actual *in situ* case, but remained simple enough to allow them to be adequately modeled. The data gained for each of the schemes agreed with expectations. In the case of phenol, both the rate and extent of biodegradation decreased with the increase in the complexity of the soil systems in the experimental schemes. The amount of phenol degraded in the wafer reaction was less than in the slurry reactor primarily due to lower water content and mass transfer rates, since phenol did not partition significantly into the soil phase. In the case of the porous tube reactor, the amount of phenol degraded was even lower primarily due to oxygen diffusion limitations. In the case of naphthalene, all three reactors produced comparable cumulative oxygen uptakes. This was mainly due to the fact that naphthalene partitioned mainly into soil organic matter, being a strongly hydrophobic compound and biodegradation of napthalene occurred mainly in the soil phase. The aqueous concentration of naphthalene was so small that the contribution of biodegradation in the aqueous phase was negligible. Modeling procedures applied to the three experimental schemes proved useful for determining biokinetic parameters for degradation of phenol. Model predictions agreed very well with experimental data. Further, the model parameters were useful in simulating treatment endpoints for these types of soil reactors. The application of this protocol to other chemicals is feasible with only minor alterations in methodology (Tabak et al., 1997).

3.5. Bioremediation of Petroleum Products

Petroleum industry is one of the major economic sectors in the world. Bioremediation of pollution caused by petroleum industry has been considered as an important environmental topical study worldwide. The most widely distributed environmental pollution can be attributed to contamination with petroleum hydrocarbons, caused by tanker accidents, storage tank ruptures, pipeline leaks, and transport accidents. Bioremediation of soils contaminated with petroleum hydrocarbons has been established as an efficient, economic, versatile, and environmentally sound treatment (Singh et al., 2004; Van et al., 2003)

Laboratory incubations of aquifer material or enrichments derived from aquifer material as well as geochemical data have suggested that, under the appropriate conditions, BTEX components of petroleum (benzene, toluene, ethylbenzene and xylene) can be degraded in the absence of molecular oxygen with either Fe (III), sulfate, or nitrate serving as the electron acceptor. BTEX degradation under methanogenic conditions has also been observed. However, especially for benzene, the BTEX contaminant of greatest concern, anaerobic degradation is often difficult to establish and maintain in laboratory incubations. Although studies to date have suggested that naturally occurring anaerobic BTEX degradation has the potential to remove significant quantities of BTEX from petroleum-contaminated aquifers, and mechanisms for stimulating anaerobic BTEX degradation in laboratory incubations have been developed, further study of the organisms involved in this metabolism and the factors controlling their distribution and activity are required before it will be possible to design rational strategies for accelerating anaerobic BTEX degradation in contaminated aquifers(Lovley,1997). Composting of horse manure together with oil wastes was found to be a suitable environment to achieve a microbial oil breakdown. The type of oil waste influenced the rate of carbon decomposition in composts, whereas the residual amounts of oils were similarly low in the four composts despite a lag-phase at initial. Oil concentrations of 7%. Repeated additions of horse manure to the composts enhanced oil decomposition. The extent of oil decomposition was higher than of organic carbon in the composts. Oil evaporation did not contribute significantly to its disappearance (Holger Kirchmann and Wasiyhun Ewnetu, 1998).

Two hydrocarbon-degrading bacterial isolates, an *Arthrobacter* sp. and a Gram-negative bacillus isolated from Kuwait oil lakes, exhibited considerable cell-surface hydrophobicity without production of exopolysaccharides in complex media. However, the bacteria produced copious amounts of exopolysaccharides in a low nutrient medium. When incubated with sawdust, Styrofoam or wheat bran, as carriers, under low nutrient conditions, stable exopolysaccharide-mediated immobilized cultures were formed. Such immobilized cultures when air-dried at room temperature survived storage for 6 weeks at 45 °C and still retained the ability to degrade hydrocarbons. Viability was retained by the immobilized *Arthrobacter* sp. and the Gram-negative bacterium at 45 °C storage for up to 6 and 12 months, respectively(Obuekwe and Esraa,2001).

The effects of biosurfactants on the biodegradation of petroleum compounds were investigated. Candida antarctica T-34 could produce extracellular biosurfactant mannosylerythritol lipids (MELs) when it was cultured in vegetable oil. In addition, in our previous study, it was found that this strain could also produce a new type of biosurfactant while it grew on n-undecane ($C_{11}H_{24}$), and the biosurfactant was named as BS-UC. In flask culture of Candida antarctica, the addition of BS-UC could improve the biodegradation rate of some n-alkanes (e.g. 90.2% for n-decane, 90.2% for n-undecane, 89.0% for dodecane), a mixture of n-alkanes (82.3%) and kerosene (72.5%). By comparing the effects of the biosurfactants BS-UC and MEL and chemical surfactants on the biodegradation of crude oil, it was found that biosurfactants could be used to enhance the degradation of petroleum compounds instead of chemical surfactants. In a laboratory scale immobilized bioreactor, the addition of biosurfactant improved not only the emulsification of kerosene in simulated wastewater but also its biodegradation rate. The highest degradation rate of kerosene by addition of MEL and BS-UC reached 87 and 90% at 15 h, respectively. The results showed that the biosurfactant BS-UC was highly promising for work on biodegradation of hydrophobic contaminants (Zhao et al., 2004).

The Indigenous bacteria, *Stenotrophomonas multophilia*, were isolated from the contaminated sites near to Jordan Petroleum Refinery and used further in the bioremediation experiments. First order kinetic equation has been proven to satisfactorily describe the biodegradation of petroleum contained in soil in the presence of the isolated bacteria. The results also showed that the first order kinetic constants for the different bioreactors vary between 0.041 and 0.0071/day. The overall kinetic constant k′ was determined based on food-to-microorganisms ratio and found to be 0.02/day (Bassim et al., 2008).

Microbiological and chemical analyses and a suitable bioreactor design were very useful for suggesting the best ways to improve biodegradation extents in a diesel-enriched soil. Samples taken in polluted soil and water revealed that bacteria from the genus *Acinetobacter* were predominant. In plate studies, *Acinetobacter* colonies produced a whitish substance with the characteristics of a biosurfactant. Remarkably, the presence of this product was evident at the field site, both in the riverbanks and in the physical recovery plant. The study of the similarities between laboratory results and the diesel spill site strongly suggested that natural conditions at the field site allowed the implementation of *in situ* bioremediation after physical removal of LNAPL (light nonaqueous-phase liquids) (Jose et al., 2001).

Hwang et al. (2001) investigated the bioremediation of diesel-contaminated using composting techniques. They found that the average first order kinetic rate constant of diesel oil was 0.099/day. Simoncyril (2001) suggested that inoculation of *Pseudomonas putida* (PP) with addition of appropriate inorganic nutrients may be a suitable method for a rapid rehabilitation of agricultural land upon pollution with crude petroleum. Odd et al. (2006) investigated biodegradation of hydrocarbons in thin oil films in seawater at low temperatures, 0 and 5 . Heterotrophic (HM) or oil-degrading (ODM) microorganisms enriched at the two temperatures showed 16S rRNA sequence similarities to several bacteria of Arctic or Antarctic origin. Biodegradation experiments were conducted with a crude mineral oil immobilized as thin films on hydrophobic Fluortex adsorbents in nutrient-enriched or sterile seawater. Chemical and respirometric analysis of hydrocarbon depletion showed that naphthalene and other small aromatic hydrocarbons (HCs) were primarily biodegraded after dissolution to the water phase, while biodegradation of larger polyaromatic hydrocarbons (PAH) and C_{10}–C_{36} n-alkanes, including n-hexadecane, was associated primarily with the oil films. Biodegradation of PAH and n-alkanes was significant at both 0 and 5 , but was decreased for several compounds at the lower temperature. n-Hexadecane biodegradation at the two temperatures was comparable at the end of the experiments, but was delayed at 0 . Investigations of bacterial communities in seawater and on adsorbents by PCR amplification of 16S rRNA gene fragments and DGGE analysis indicated that predominant bacteria in the seawater gradually adhered to the oil-coated adsorbents during biodegradation at both temperatures. Sequence analysis of most DGGE bands aligned to members of the phyla *Proteobacteria* (*Gammaproteobacteria*) or *Bacteroidetes*. Most sequences from experiments at 0 revealed affiliations to members of Arctic or Antarctic consortia, while no such homology was detected for sequences from degradation experiment run at 5 . In conclusion, marine microbial communities from cold seawater have potentials for oil film HC degradation at temperatures ≤5 , and *psychrotrophic* or *psychrophilic* bacteria may play an important role during oil HC biodegradation in seawater close to freezing point. Obbard et al. (2004) reported on the ability of crude palm oil (CPO) and several component fatty acids to stimulate the indigenous microbial biomass and degrade selected petroleum hydrocarbons in oil-spiked beach sediments under laboratory conditions as a prelude to further field trials in Singapore. Amendment of simple organic carbon, in the presence of inorganic nutrients, to oil contaminated beach sediments can potentially stimulate the biodegradation of hydrocarbons by the indigenous microbial biomass. The ability of crude palm oil (CPO) and fatty acids, in the presence of soluble inorganic nutrients (C:N:P = 100:10:1), to stimulate biodegradation in sediments amended with Arabian light crude oil was investigated in laboratory microcosms over a 30-day period. The presence of nutrients alone enhanced the most probable number (MPN) of hydrocarbon degrading bacteria by almost 12 times after 30 days, and by 26 times in sediment also amended with CPO, thereby confirming its ability to stimulate the biomass. Addition of individual fatty acids to nutrients-amended sediments resulted in MPN enhancements of up to 170 times, where straight alkanes (i.e., C_{17} and C_{18}) were completely degraded. Amendment with CPO, myristic, oleic, linoleic and palmitic acids (0.5% dry weight equivalent) also enhanced the metabolic activity of the entire microbial biomass, as measured by intracellular dehydrogenase activity, reaching a maximum of 30 μg INTF g dry sed^{-1} h^{-1}. Loss of branched alkanes was significantly enhanced, where loss of pristane was complete and phytane loss reached 98% in sediment treated with myristic acid. This work highlights the considerable benefit of adding readily oxidizable fatty acids for enhancing *in situ* bioremediation of hydrocarbons in nutrient-amended beach sediments.

A combination of geochemical, microbiological and isotopic methods were used to evaluate in-situ bioremediation of petroleum hydrocarbons at one site contaminated with refinery waste and a second site contaminated with aviation gasoline at Alameda Point, California. At each site, geochemical and microbiological characteristics from four locations in the contaminated zone were compared to those from two uncontaminated background locations. At both sites, the geochemical indicators of in-situ biodegradation included depleted soil gas and groundwater oxygen, elevated groundwater alkalinity, and elevated soil gas carbon dioxide and methane in the contaminated zone relative to the background. Radiocarbon content of methane and carbon dioxide measured in soil gas at both sites indicated that they were derived from hydrocarbon contaminant degradation. Direct microscopy of soil core samples using cell wall stains and activity stains, revealed elevated microbial numbers and enhanced microbial activities in contaminated areas relative to background areas, corroborating geochemical findings. While microbial plate counts and microcosm studies using soil core samples provided laboratory evidence for the presence of some microbial activity and contaminant degradation abilities, they did not correlate well with either contaminant location, geochemical, isotopic, or direct microscopy data (Vishvesh et al., 2002).

A field-scale study was conducted by Sanjeet et al. (2001) in a 4000 m^2 plot of land contaminated with an oily sludge by use of a carrier-based hydrocarbon-degrading bacterial consortium for bioremediation. The land belonged to an oil refinery. The site selected for bioremediation contained approximately 300 tons of oily sludge. The plot was divided into four blocks, based on the extent of contamination. Blocks A, B, and C were treated with the bacterial consortium, whereas Block D was maintained as an untreated control. In Block A, at time zero, i.e., at the beginning of the experiment, the soil contained as much as 99.2 g/kg of total petroleum hydrocarbon (TPH). The application of a bacterial consortium (1 kg carrier-based bacterial consortium/10 m^2 area) and nutrients degraded 90.2% of the TPH in 120 days, whereas in block D only 16.8% of the TPH was degraded. This study validates the large-scale use of a carrier-based bacterial consortium and nutrients for the treatment of land contaminated with oily sludge, a hazardous hydrocarbon waste generated by petroleum industry.

In-situ bioremediation of petroleum waste sludge in landfarming sites of Motor Oil Hellas (petroleum refinery) was studied by monitoring the changes of the petroleum composition of the waste sludge, as well as the changes in the structure of the microbial community, for a time period of 14 months. The analyses indicated an enhanced degradation of the petroleum hydrocarbons in the landfarming areas. A depletion of n-alkanes of approximately 75–100% was obtained. Marked changes of the microbial communities of the landfarms occurred concomitantly with the degradation of the petroleum hydrocarbons. The results obtained from terminal restriction fragment length polymorphism (T-RFLP) analysis of polymerase chain reaction (PCR) amplified 16S rRNA genes demonstrated that bacteria originating from the refinery waste sludge and newly selected bacteria dominated the soil bacterial community during the period of the highest degradation activity. However, the diversity of the microbial community was decreased with increased degradation of the petroleum hydrocarbons contained in the landfarms. T-RFLP fingerprints of bacteria of the genera Enterobacter and Ochrobactrum were detected in the landfarmed soil over the entire treatment period of 14 months. In contrast, the genus Alcaligenes appeared in significant numbers only within the 10 month old landfarmed soil. Genes encoding catechol 2,3-dioxygenase (subfamily I.2.A) were detected only in DNA of the untreated refinery waste sludge. However, none of the genes known to encode the enzymes alkane hydroxylase AlkB, catechol 2,3-dioxygenase (subfamily I.2.A) and naphthalene dioxygenase nahAc could be detected in DNA of the landfarmed soils (Katsivela et al., 2005).

The effects of bioremediation program of sterile agricultural soils contaminated with crude petroleum were determined with a view to developing a suitable technique for rehabilitation of similar environments upon pollution by oil spillage. Sterile soils inoculated with the soil bacterium, *Pseudomonas putida* (PP), with inorganic nutrients monitoring and supplementation constituted the experimental set-ups (ESU). The control set-ups (CSU) contained all the materials present in ESU except that they were not inoculated with PP. In ESU at week 9, the oil pollutant was completely biodegraded, and the inorganic nutrient ions, particularly PO4 and NO^3, were significantly utilized. In contrast, there were no significant changes in the concentrations of oil and inorganic nutrients in CSU. Also, the percentage germination and growth profiles of cress seeds (*Lepidium* sp.) planted as evidence of the recovery of the oil-impacted soils were poor in CSU (27.5%) with pronounced abnormal morphology when compared with the results obtained for ESU (98.8%). Inoculation of PP with addition of appropriate inorganic nutrients may be a suitable method for a rapid rehabilitation of agricultural land upon pollution with crude petroleum (Simoncyril, 2001). The possibility of quantifying in situ levels of transcripts for dissimilatory (bi) sulfite reductase (dsr) genes to track the activity of sulfate-reducing microorganisms in petroleum-contaminated marine harbor sediments was evaluated. Phylogenetic analysis of the cDNA generated from mRNA for a ca. 1.4 kbp portion of the contiguous *dsrA* and *dsrB* genes suggested that Desulfosarcina species, closely related to cultures known to anaerobically oxidize aromatic hydrocarbons, were active sulfate reducers in the sediments. The levels of *dsrA* transcripts (per μg total mRNA) were quantified in sediments incubated anaerobically at the in situ temperature as well as in sediments incubated at higher temperatures and/or with added acetate to increase the rate of sulfate reduction. Levels of *dsrA* transcripts were low when there was no sulfate reduction because the sediments were depleted of sulfate or if sulfate reduction was inhibited with added molybdate. There was a direct correlation between *dsrA* transcript levels and rates of sulfate reduction when sulfate was at ca. 10 mM in the various sediment treatments, but it was also apparent that within a given sediment, *dsrA* levels increased over time as long as sulfate was available, even when sulfate reduction rates did not increase. These results suggest that phylogenetic analysis of dsr transcript sequences may provide insight into the active sulfate reducers in marine sediments and that quantifying levels of *dsrA* transcripts can indicate whether sulfate reducers are active in particular sediment. Furthermore, it may only be possible to use *dsrA* transcript levels to compare the relative rates of sulfate reduction in sediments when sulfate concentrations, and possibly other environmental conditions, are comparable (Kuk-Jeong Chin et al., 2008).

CONCLUSION

Biotechnology, from a global perspective, can be expected, within the next 20 years, to transform the sectors of health, pharmaceuticals, agriculture, food and the environment. As a tool and a social process, biotechnology has influenced the development of societies and the advancement of civilization. Located and rooted within national borders, its productive, scientific and commercial aspects are generally a function of the multinationals. Sensitive to the dictates of global markets biotechnology (Callan, 1997) has a transformative capacity in fostering and driving the expansion of the world's economy. It permeates national and regional frontiers and, is at the basis of new market-oriented processes and products (Oman, 1994). Globalization of markets involves an unavoidable interdependency of the world's economies, increasing competition and participation in cross border financing and transactions and a concern for the protection of intellectual property. Advances in environmental health science and technology can be exploited to prevent long-term disaster and irreparable damage due to environmental negligence. The harmful effects of chemicals can be prevented or at least minimized. There is an urgent need for governments and scientists to resolve these emerging problems and to plan for future prevention (DaSilva, 1998). A number of bioremediation strategies for world water resources, soils and petroleum products have been developed to degrade highly stable, toxic matter. With the help of advances in bioinformatics, biotechnology holds a bright future for developing bioprocesses for environmental applications. However, there are still many pollutants persist in the environment.

REFERENCES

Adhya TK (2008) Biodegradation and Bioremediation. *Indian J. Microbiol.* 48:1-2.

Akhond MAY, Machray GC (2008) Biotech crops: technologies, achievements and prospects. *Euphytica.* DOI 10.1007/s10681-008-9823-1

Allen KN (1991) Seasonal variation of selenium in outdoor experimental stream-wetland systems. *Journal of Environmental Quality.* 20: 865–868.

Anupa Nair, Asha A Juwarkar, Sanjeev K Singh (2007) Production and Characterization of Siderophores and its Application in Arsenic Removal from Contaminated Soil. *Water, Air, & Soil Pollution.* 180:199–212.

Archana Chauhan, Fazlurrahman, John G. Oakeshott, Rakesh K. Jain (2008) Bacterial metabolism of polycyclic aromatic hydrocarbons:strategies for bioremediation. *Indian J. Microbiol.* 48:95–113.

Arntzen CJ, Mason HS, Haq TA (2001) Oral immunization with transgenic plants. US Patent 6,194,560.

Basheer S, Kut OM, Prenosil JE, Bourne JR (1992) Kinetics of enzymatic degradation of cyanide. *Biotechnol. Bioeng.* 39:629–634.

Bassim E Abbassi, Walid D Shquirat (2008) Kinetics of Indigenous Isolated Bacteria used for Ex-Situ Bioremediation of Petroleum Contaminated Soil. *Water, Air, & Soil Pollution.* 192:221–226.

Beachy RN (1997) Mechanisms and applications of pathogenderived resistance in transgenic plants. *Curr. Opin. Biotechnol.* 8:215–220.

Brunello Ceccanti, Grazia Masciandaro, Carlos Garcia, Cristina Macci, Serena Doni (2006) Soil bioremediation: combination of earthworms and compost for the ecological remediation of a hydrocarbon polluted soil. *Water, Air, & Soil Pollutio.* 177: 383–397.

Callan, B (1997) Why production technology is not a measure of competitiveness in the biotechnologies. *Science and Public Policy.* 24, 146-160.

Camp WV (2005) Yield enhancement genes: seeds for growth. *Curr. Opin. Biotechnol.* 16:147–153.

Chen Z, Huang GH, Chan CW, Geng LQ, Xia J (2003) Development of an expert system for the remediation of petroleum-contaminated sites. *Environmental Modeling and Assessment.*.8:323–334.

Cohen JI, Quemada H, Frederick R (2003) Food safety and GM crops: implications for developing-country research. In: Food safety in food security and food trade. Focus 10, Brief 16 of 17. International Food Policy Research Institute, Washington, DC, USA.

Comeau Y, Hall K, Hancock R, Oldham W (1986) Biochemical model for enhancing biological phosphorus removal. *Water Res.* 20(2):1511–1521.

DaSilva E (1998) Review: Biotechnology: developing countries and globalization. World *Journal of Microbiology & Biotechnology.* 14, 463-486

De Block M, Botterman J, Vandeweile M, Dockx J, Thoen C, Gossele V, Movva NR, Thompson C, van Montagu M, Leemans J (1987) Engineering herbicide resistance in plants by expression of a detoxyfying enzyme. *EMBO J.* 6:2513–2518

Dierberg FF, DeBusk TA, Goulet Ja NA (1987) Removal of copper and lead using a thin film technique, In: Reddy KB, Smith WH(eds). Aquatic plants for water treatment and resource recovery. Magnolia Publishing Inc, Florida.

Dileep K Singh (2008) Biodegradation and bioremediation of pesticide in soil: concept, method and recent developments. *Indian J. Microbiol.* 48:35–40.

Doty SL, Shang TO, Wilson AM, Tangen J, Westergreen AD, Newman LA, Strand SE, Gordan MP (2000) Enhanced metabolism of halogenated hydrocarbons in transgenic plants containing mammalian cytochrome P450 2E1. *Proc. Natl. Acad. Sci. U.S.A.* 97:6287–6291

Duan J, Gregory J (2003) Coagulation by hydrolysing metal salts. *Adv. Coll. Inter. Sci.* 100–102: 475-502.

Duke SO (ed) (1996) Herbicide-resistant crops: agricultural, environmental, economic, regulatory, and technical aspects. CRC Press, Boca Raton

Dunwell JM (1999) Transgenic crop plants: the next generation or an example of 2020 vision. *Ann. Bot.* 84: 269–277.

Ebbs S (2004) Biological degradation of cyanide compounds. *Environ. Biotechnol.* 15:231-236.

EFSA [European Food Safety Authority] (2004) Guidance document of the GMO Panel for the risk assessment of genetically modified plants and derived food and feed (http://www.efsa.eu.int/science/gmo/gmo_guidance/ 660_en.html)

EPA (Environmental Protection Agency) (2002) EPA's Regulation of biotechnology for use in pest management (735-F-02-003). Available free at: http://www.epa.gov/pesticides/ biopesticides/reg_og_biotech/epategofbiotech.htm.

EPA (Environmental Protection Agency) (2003) Event MON863 Bt Cry3Bb1 corn biopesticide registration action document. Available free at: http:www.epa.gov/pesticides/ biopesticides/ingredients/tech_docs/brad_006484.htm.

Eric D van Hullebusch, Piet NL Lens, Henry H Tabak (2005) Developments in bioremediation of soils and sediments polluted with metals and radionuclides. 3. Influence of chemical speciation and bioavailability on contaminants immobilization/mobilization bio-processes. *Reviews in Environmental Science & Biotechnology.* 4:185–212.

Eugenio Foresti, Marcelo Zaiat, Marcus Vallero (2006) Anaerobic processes as the core technology for sustainable domestic wastewater treatment: Consolidated applications, new trends, perspectives, and challenges. *Reviews in Environmental Science and Biotechnology.* 5:3-19.

Ezzi M, Lynch JM (2005) Biodegradation of cyanide by Trichoderma spp. and Fusarium spp. *Enz. Microb. Technol.* 36:849 -854.

Fan TWM, Lane AN, Higashi RM (1997) Selenium biotransformations by a euryhaline microalga isolated from a saline evaporation pond. *Environmental Science & Technology.* 31:569-576.

Faull JL, Graeme-Cook KA, Pilkington BL (1994) Production of an isonitrile antibiotic by an UV-induced 14 mutant of Trichoderma harzianum. *Phytochemistry.* 36:1273-1276.

Ferry N, Edwards MG, Gatehouse JA, Gatehouse AMR (2004) Plant-insect interactions: molecular approaches to insect resistance. *Curr. Opin. Biotechnol.* 15:155–161.

Figueira MM, Ciminelli VST, Linardi VR (1995) Bacterial degradation of metal cyanide complexes. In: Jerez C.A.,Vargas T., Toledo H., Wiertz J.V. (eds), Biohydrometallurgical Processing, University of Chile, Chile, pp. 333-339.

Garcia S, Ezcurra E, Schoel B, Acevedo F, Soberon J, Snow AA (2005) Absence of detectable transgenes in local landraces of maize in Oaxaca, Mexico. *Proc. Natl. Acad. Sci. U.S.A.* 35:12338–12343

George P Prpich, Rachel L Adams, Andrew J Daugulis (2006) Ex situ bioremediation of phenol contaminated soil using polymer beads. *Biotechnol. Lett.* 28:2027-2031.

Gerhard Wenzel(2006) Molecular plant breeding: achievements in green biotechnology and future perspectives. *Appl. Microbiol. Biotechnol.* 70: 642–648

Gilbert ES, Crowley DE (1997) Plant compounds that induce polychlorinated biphenyl biodegradation by *Arthrobacter* sp. strain B1B. *Appl. Environ. Microbiol.* 63:1933–1938.

Gressel J (1996) The potential roles for herbicide-resistant crops in world agriculture. In: Duke SO (ed) Herbicide-resistant crops. CRC Press, Boca Raton, pp 231–250

Hansen HCB, Koch CB, Nancke-Krogh H, Borggaard OK, Sorensen J (1996) Abiotic nitrate reduction to ammonium: key role of green rust. *Environmental Science & Technology.* 30:2053–2056.

Heck GR, CaJacob CA, Padgette SR (2003) Discovery, development, and commercialization of Roundup Ready_ crops. In: Vasil IK (ed) Plant biotechnology 2002 and beyond. Kluwer, Dordrecht, pp 139–142

Henry H Tabak, Piet Lens, Eric D van Hullebusch, Winnie Dejonghe (2005) Developments in bioremediation of soils and sediments polluted with metals and radionuclides - 1. Microbial processes and mechanisms affecting bioremediation of metal contamination and influencing metal toxicity and transport. *Reviews in Environmental Science and Biotechnology.* 4:115-156.

Herbers K, Sonnewald U (1999) Production of new/modified proteins in transgenic plants. *Curr. Opin. Biotechnol.* 10:163–168.

Hohn B, Puchta H (2003) Some like it sticky: targeting of the rice gene *Waxy*. *Trends Plant. Sci.* 8:51–53.

Holger Kirchmann, Wasiyhun Ewnetu (1998) Biodegradation of petroleum-based oil wastes through composting. *Biodegradation.* 9:151-156.

Huang XD, Alawi YE, Penrose DM, Glick BR, Greenberg BM (2004) A multi-process phytoremediation system for removal of polycyclic aromatic hydrocarbons from contaminated soils. *Environ. Pollut.* 130:465-476.

Hwang EY, Wan N, Park JS (2001) Recycling of remediated soil for effective composting of dieselcontaminated soil. *Compost. Science and Utilization.* 9 (2), 143-149.

Jain SK, Vasudevan P, Jha NK (1989) Removal of some heavy metals from polluted waters by aquatic plants: studies on duckweed and water velvet. *Bio Wastes.* 28:115-126.

James C (2002) Global GM Crop Area Continues to Grow and Exceeds 50 million Hectares for First Time in 2001. ISAAA. Activities. In: http / /www.isaaa.org/ pres release /Global Area- Jan2002.htm

James C (2007) Global status of commercialized biotech/GM crops: 2007. ISAAA Brief 37

Joanne Baxter, Stephen P. Cummings (2006) The current and future applications of microorganism in the bioremediation of cyanide contamination. Antonie van Leeuwenhoek, 90:1-17.

Jos´e LR Gallego, Jorge Loredo, Juan F Llamas, Fernando V´azquez, Jes´us S´anchez (2001) Bioremediation of diesel-contaminated soils: Evaluation of potential *in situ* techniques by study of bacterial degradation. *Biodegradation.* 12: 325-335.

Joshi L, Lopez LC (2005) Bioprospecting in plants for engineered proteins. *Curr. Opin. Plant. Biol.* 8:223–226.

Ka′ tia Maria Gomes Machado, Da′ cio Roberto Matheus, Regina Teresa Rosim Monteiro, Vera Lu′ cia Ramos Bononi (2005) Biodegradation of pentachorophenol by tropical basidiomycetes in soils contaminated with industrial residues. *World Journal of Microbiology & Biotechnology.* 21:297-301.

Kara Y (2004) Bioaccumulation of Copper from Contaminated Wastewater by Using Lemna minor. *Bull. Environ. Contam. Toxicol.* 72:467-471.

Katsivela E, Moore ERB, Maroukli D, Strompl C, Pieper D, Kalogerakis N (2005) Bacterial community dynamics during in-situ bioremediation of petroleum waste sludge in landfarming sites. *Biodegradation.* 16:169–180.

Keiko Yonekura-Sakakibara, Kazuki Saito(2006) Review: genetically modified plants for the promotion of human health. *Biotechnol. Lett.* 28:1983–1988

Kelley BD (2001) Biochemical engineering: bioprocessing of therapeutic proteins. *Curr. Opin. Biotechnol.* 12:173–174.

Kjeldsen P (1999) Behaviour of cyanides in soil and groundwater: a review. Water, Air, & Soil Pollution. 115:279–308.

Krämer U (2005) Phytoremediation: novel approaches to cleaning up polluted soils. *Curr. Opin. Biotechnol.* 16:133–141.

Krishnani KK, Gupta BP, Joseph KO, Muralidhar M, Sarda C, Nagavel A, Parimala V (2003) Decontamination of nitrogenous toxicants from brackish water using plant and animal extracts. *Bull. Environ. Contam. Toxicol.* 71:196–203.

Kuk-Jeong Chin, Manju L Sharma, Lyndsey A Russell, Kathleen R O'Neill, Derek R Lovley (2008) Quantifying Expression of a Dissimilatory (bi) Sulfite Reductase Gene in Petroleum-Contaminated Marine Harbor Sediments. *Microb. Ecol.* 55:489–499.

Kumada Y, Kisa T, Kuwano Y, Koshikawa H, Tashiro,Yasukagawa E, Shimizu Y (2005) Microbial community analysis of PAHs degradation in soil. Proc 5th Korea–Japan joint seminar on geo-environmental engineering, 4th July, Seoul National University, Seoul.

Lal R, Dogra C, Malhotra S, Sharma P, Pal R (2006) Diversity, Distribution and Divergence of *lin* genes in hexachlorocyclohexane degrading sphingomonads. *Trends in Biotechnology.* 24:121–130.

Lee D, Natesan E (2006) Evaluating genetic containment strategies for transgenic plants. - *Trends Biotechnol.* 24(3):109– 114.

Letourneau, DK, Burrows BE(2002) Genetically Engineered Organisms. Assessing Environmental and Human Health Effects. CRC Press, Boca Raton.

Lettinga G (1995) Anaerobic digestion and wastewater treatment systems. *Antonie van Leeuwenhoek.* 67:3–28.

Lettinga G, Roersma R, Grin P, de Zeew W, Hulshof Pol LL,van Velsen L, Hobma S, Zeeman G. (1982) Anaerobic treatment of low strength waste waters. In: Hughes E, Stafford DA, Wheatley BI, Baader W, Lettinga G, Nyns EJ, Verstraete W, Wentworth RL (Eds) Anaerobic Digestion 1981 (pp 271–291). Elsevier Biomedical Press BV, Amsterdam.

Lordi DT, Lue-Hing C, Whitebloom SW, Kelada N, Dennison S (1980) Cyanide problems in municipal waste water treatment plants. *J. Wat. Poll. Cont. Fed.,* 52:597 –609.

Lovley DR (1997) Potential for anaerobic bioremediation of BTEX in petroleum-contaminated Aquifers. *Journal of Industrial Microbiology & Biotechnology.* 18:75–81.

McCarty PL (1982) One hundred years of anaerobic treatment. In: Hughes DE, Stafford DA, Wheatley BI, Baader W, Lettinga G, Nyns EJ, Verstraete W, Wentworth RL(Eds) Anaerobic Digestion (pp 3–22). Elsevier Biomedical Press BV, Amsterdam.

Mohapatra, PK, Mohanty, RC (1992) Determination of water quality of two water bodies using algal bioassay method. *Phycos.* 31(1,2):77–84.

Mohit Kumar C, Vidya Lakshmi, Sunil Khanna (2008) Microbial biodiversity and in situ bioremediation of endosulfan contaminated soil. *Indian J. Microbiol.* 48:128-133.

Moose SP, Mumm RH (2008) Molecular plant breeding as the foundation for 21st century crop improvement. *Plant Physiol.* 147:969–977.

NAS (National Academy of Sciences) (2002) Environmental effects of transgenic plants: The scope and adequacy of regulation. Washington, DC: National Academy Press. Available free at http://books.nap.edu/books/0309082633/html/index/html.

NAS (National Academy of Sciences)(2000) Genetically modified pest-protected plants: Science and regulation. Washington, DC: National Academy Press. Available free at http://www.napedu/books/0309069300/html/.

Niu QW, Lin SS, Reyes JL, Chen KC, Wu HW, Yeh SD (2006) Expression of artiWcial microRNAs in transgenic *Arabidopsis thaliana* confers virus resistance. Nat. Biotechnol. 24(11):1420–1428.

Nordlee JA, Taylor SL, Townsend JA, Thomas LA, Bush RK (1996) Identification of a Brazil-nut allergen in transgenic soybeans. *New Eng. J. Med.* 334(11): 688–692.

NRC (National Research Council) (2001) Environmental Effects of Transgenic Plants, 320 pp. National Academy Press,Washington, DC.

Obbard JP, NG KL, XU R. (2004) Bioremediation of petroleum contaminated beach sediments: use of crude palm oil and fatty acids to enhance indigenous biodegradation. *Water, Air, and Soil Pollution. 1*57:149–161.

Obuekwe CO, Esraa M Al-Muttawa (2001) Self-immobilized bacterial cultures with potential for application as ready-to-use seeds for petroleum bioremediation. *Biotechnology Letters.* 23: 1025–1032.

Odd G Brakstad, Kristin Bonaunet (2006) Biodegradation of petroleum hydrocarbons in seawater at low temperatures (0-5) and bacterial communities associated with degradation. *Biodegradation.* 17:71-82.

Ohtsubo Y, Shimura M, Delawary M, Kimbara K, Takagi M, Kudo T, Ohta A, Nagata Y (2003) Novel approach to the improvement of biphenyl and polychlorinated biphenyl degradation activity: promoter implantation by homologous recombination. *Appl. Environ. Microbiol.* 69:146-153.

Oman, C (1994) Globalization and Regionalization - The Challenge for Developing Countries. Paris; Organization for Economic Co-operation and Development.

Padgette SR, Re DB, Barry GF, Eichholtz DE, Delannay X, Fuchs RL, Kishore GM, Fraley RT (1996) New weed control opportunities: development of soybeans with a Roundup Ready gene. In: Duke SO (ed) Herbicide-resistant crops. CRC Press, Boca Raton, pp 53–84

Parimala V, Krishnani KK, Gupta BP, Jayanthi M, Abraham M (2004) Phytoremediation of chromium from seawater using five different products from coconut husk. *Bull. Environ. Contam. Toxicol.* 73:31–37.

Parker WL, Rathnum ML, Johnson JH, Wells JS, Principe PA, Sykes RB (1988) Aerocyanidin, a new antibiotic produced by Chromobacterium violaceum. *J. Antibiot,* 41:454 –460.

Rai UN, Dwivedi S, Tripathi RD, Shukla OP, Singh NK (2005) Algal biomass: An economical method for removal of chromium from tannery effluent. *Bulletin of Environmental Contamination and Toxicology.* 75, 297– 303.

Rai UN, Smita Dubey, Shukla OP, Dwivedi S, Tripathi RD (2008) Screening and identification of early warning algal species for metal contamination in fresh water bodies polluted from point and non-point sources. *Environ. Monit. Assess.* 144:469–481.

Regine Eibl , Dieter Eibl (2008) Design of bioreactors suitable for plant cell and tissue cultures. *Phytochem. Rev.* 7:593–598.

Rissler J, Mellon M (1993) Perils amidst the promise: ecological risks of transgenic crops in a global market. Union of Concerned Scientists, Cambridge, USA

Rissler J, Mellon M (1996) The Ecological Risks of Engineered Organisms. MIT Press, Cambridge, MA.

Ritzenthaler C (2005) Resistance to plant viruses: old issue, new answers? *Curr. Opin. Biotechnol.* 16:118–122.

Rommens C, Kishore GM (2000) Exploiting full potential of disease-resistance genes for agricultural use. *Curr. Opin. Biotechnol.* 11:120–125.

Sanjeet Mishra, Jeevan Jyot, Ramesh Chander Kuhad, Banwari Lal (2001) In Situ Bioremediation Potential of an Oily Sludge-Degrading Bacterial Consortium. *Current Microbiology.* 43:328–335.

Santiago Villaverde (2004) Recent developments on biological nutrient removal processes for wastewater treatment. *Reviews in Environmental Science and Bio/Technology.* 3:171–183.

Sarkar AD, Ray Amulya N, Shrivastava Subhodeep Sarker (2006) Molecular Biomarkers: Their significance and application in marine pollution monitoring. *Ecotoxicology.* 15:333–340.

Schack-Kirchner H, Hildebrand EE (1998). Changes in soil structure and aeration due to liming and acid irrigation. *Plant and Soil.* 199:167–176.

Scheller J, Conrad U (2005) Plant-based material, protein and biodegradable plastic. *Curr. Opin. Plant. Biol.* 8:188–196

Schoeman JJ, Steyn A (2003) Nitrate removal with reverse osmosis in a rural area in South Africa. *Desalination.* 155: 15–26.

Sharma HC, Crouch JH, Sharma KK, Seetharama N, Hash CT (2002) Applications of biotechnology for crop improvement: prospects and constraints. *Plant Sci.* 163:381–395.

Siddique Tariq, Benedict C Okeke, Arshad Muhammad, Frankenberger William T Jr (2003) Enrichment and Isolation of Endosulfan-Degrading Microorganisms. *J. Environ. Qual.* 32:47–54.

Simoncyril U Nwachukwu (2001) Bioremediation of Sterile Agricultural Soils Polluted with Crude Petroleum by Application of the Soil Bacterium, *Pseudomonas putida*, with Inorganic Nutrient Supplementations. *Current Microbiology.* 42:231–236.

Singer AC, Jury W, Luepromchai E, Yahng CS, Crowley DE (2001) Contribution of earthworms to PCB bioremediation. *Soil-Biol. Biochem.* 33:765–776.

Singh A, Ward OP (2004) Applied Bioremediation and Phytoremediation.Soil Biology, Vol. 1. Springer Verlag, Berlin, 281.

Snow AA, Moran-Palma P (1997) Commercialization of transgenic plants: potential ecological risks. *BioScience.* 47: 86–96.

Song J, Bradeen JM, Naess SK, Raasch JA, Wielgus SM, Haberlach GT (2003) Gene RB cloned from *Solanum bulbocastanum* confers broad spectrum resistance to potato late blight. *Proc. Natl. Acad. Sci. U.S.A.* 100(16):9128–9133.

Soo Cho J, Wilson JT, DiGiulio DC, Vardy JA, Choi W (1997) Implementation of natural attenuation at a JP-4 jet fuel release after active remediation. *Biodegradation.* 8:265–273.

Stoger E, Ma JKC, Fischer R, Christou P (2005) Sowing the seeds of success: pharmaceutical proteins from plants. *Curr. Opin. Biotechnol.* 16:167–173

Susarla S, Medina VF, McCutcheon SC (2002) Phytoremediation: an ecological solution to organic chemical contamination. *Ecol. Eng.* 18:647–658.

Sutherland TD, Horne I, Harcourt RL, Russel RJ, Oakeshott JG (2002) Isolation and characterization of a *Mycobacterium* strain that metabolizes the insecticide endosulfan. *J. Appl. Microbiol.* 93:380–389.

Sutherland TD, Horne I, Lacey MJ, Harcourt RL, Russell RJ, Oakeshott JG (2000) Enrichment of an endosulfan-degrading mixed bacterial culture. *Applied and Environmental Microbiology.* 66:2822–2828.

Svetlozar Velizarov, Joao G Crespo, Maria A Reis (2004) Removal of inorganic anions from drinking water supplies by membrane bio/processes. *Reviews in Environmental Science & Biotechnology.* 3: 361-380.

Tabak HH, Govind R (1997) Protocol for determining bioavailability and biokinetics of organic pollutants in dispersed, compacted and intact soil systems to enhance in situ bioremediation. *Journal of Industrial Microbiology & Biotechnology.* 18:330–339.

Terry C Hazen1, Henry H Tabak (2005) Developments in bioremediation of soils and sediments polluted with metals and radionuclides: 2. Field research on bioremediation of metals and radionuclides. *Reviews in Environmental Science and Bio/Technology.* 4:157–183.

Traynor PL, Westwood JH (1999) Ecological Effects of Pest Resistance Genes in Managed Ecosystems. *Information Systems for Biotechnology.* Blacksburg, VA.

Van Hamme JD, Singh A, Ward OP (2003) Recent advances in petroleum microbiology. *Microbiol. Mol. Biol. Rev.* 67:503-549.

Vasil IK (1996) Phosphinothricin-resistant crops. In: Duke SO (ed) Herbicide-resistant crops. CRC Press, Boca Raton, pp 85–91

Vasil IK (2003) The science and politics of plant biotechnology—a personal perspective. *Nat. Biotechnol.* 21:849–851

Vasil IK (ed) (1999) Advances in cellular and molecular biology of plants. *Molecular improvement of cereal crops.* Kluwer, Dordrecht, vol 5

Vasil IK, Bean S, Zhao J, McCluskey P, Lookhart G, Zhao H, Altpeter F, Vasil V (2001) Evaluation of baking properties and gluten protein composition of field grown transgenic wheat lines expressing high molecular weight glutenin gene 1Ax1. *J. Plant Physiol.* 158:521–528

Venkata S Mohan, Takuro Kisa, Takeru Ohkuma, Robert A Kanaly, Yoshihisa Shimizu(2006) Bioremediation technologies for treatment of PAH-contaminated soil and strategies to enhance process efficiency. *Rev. Environ. Sci. Biotechnol.* 5:347–374.

Vishvesh K Bhupathiraju, Paula Krauter, Hoi-Ying N Holman, Mark E Conrad, Paul F. (2002) Assessment of in-situ bioremediation at a refinery waste-contaminated site and an aviation gasoline contaminated site. *Biodegradation.* 13:79–90.

Weir KM, Sutherland TD, Horne I, Russell RJ, Oakeshott JG (2006) A Single Monooxygenase, Ese, Is Involved in the Metabolism of the Organochlorides Endosulfan and Endosulfate in an *Arthrobacter* sp. *Appl. Environ. Microbiol.* 72:3524–3530.

Whitton BA, Say PJ, Jupp BP (1982) Accumulation of zinc, cadmium and lead by the aquatic liverwort Scapania. *Environmental Pollution.* 3:299–316.

Wolfenbarger LL, Phifer PR (2000) The ecological risks and benefits of genetically engineered plants. *Science.* 290: 2088–2093.

Wolters V (2000) Invertebrate control of soil organic matter stability. *Biol. Fert. Soils.* 31:1–19.

Zaharie Moldovan, Romeo Chira, Alfredo C Alder (2008) Environmental exposure of pharmaceuticals and musk fragrances in the Somes River before and after upgrading the municipal wastewater treatment plant Cluj-Napoca, Romania. *Environ. Sci. Pollut. Res.* DOI 10.1007/s11356-008-0047-7.

Zhao zhe Hua, Yan Chen, Guo cheng Du, Jian Chen (2004) Effects of biosurfactants produced by *Candida antarctica* on the biodegradation of petroleum compounds. World *Journal of Microbiology & Biotechnology.* 20:25–29.

In: Environmentalism: Environmental Strategies...
Editors: E. Kahraman and A. Baig, pp.41-72 ISBN: 978-1-60741-634-0

Chapter 2

ENERGY MASTER PLANNING: THE CASE OF NEW JERSEY

Frank A. Felder, Nancy Mantell, Nora Lovrien Buehler and Andrew Cottrell
Edward J. Bloustein School of Planning and Public Policy
Rutgers, The State University of New Jersey

ABSTRACT

In December 2006, the State of New Jersey initiated its energy master planning process. The aim was to develop and then implement a comprehensive plan to address energy supply and consumption through 2020. In a state faced with high energy prices and deep-seated concerns regarding the impacts of energy use on air quality and of climate change on the New Jersey landscape, the governor called for the plan to develop aggressive greenhouse gas reduction goals. Nearly two years after the process began, the New Jersey Energy Master Plan was published in October 2008.

Getting to the final shape of the plan entailed extensive collaboration among State agencies and departments and committed stakeholder involvement from utilities, environmental groups, and citizens. The plan includes energy efficiency, renewable energy, and infrastructure and land use policies. Every sector of the economy is affected and every energy source addressed. Responsible parties, funding streams, and timelines have been developed for each policy.

Such a comprehensive plan was possible for three reasons. First, strong executive leadership kept the New Jersey Energy Master Plan at the top of the list of policy priorities. Second, a broad and inclusive process engaged key stakeholders and residents and helped build support for the plan. Third, the State engaged an academic team to develop a comprehensive modeling platform designed to analyze both the energy and macroeconomic impacts of the plan. Through this multi-faceted approach, the New Jersey Energy Master Plan has shown that investing in a cleaner energy future should be beneficial to the economy, not harmful as some might claim.

This chapter defines the concept of energy master planning and emphasizes both the benefits and the limits of the approach. It then establishes the importance of leadership and institutionalization of the plan to set the groundwork for successfully achieving the goals. This includes identifying and engaging key stakeholders in the process of developing the plan. Finally, the use of scenario modeling illustrates how alternative policy proposals are vetted for their relative impact on macroeconomic and environmental indicators for the state. This quantitative analysis—though far from conclusive—provides a level of legitimacy to the planning process and should lessen potential critiques. Involving stakeholders and experts in modeling is as important as engaging them in the development of the policy platform.

I. Energy Master Planning

A. Introduction

Energy is required for a functional, efficient, and healthy society. Systems that we rely upon everyday—transportation, electricity, water, communications—would grind to a halt without access to energy. In the event of a power outage, lights turn off, fans and freezers stop working, and traffic lights go dark. These events have significant implications for comfort, the economy, and public safety in their own right; imagine, though, adding gasoline shortages, natural gas pipeline interruptions, and water treatment plant shutdowns to the list of possible scenarios in a world of unreliable energy supplies. These could result from any number of possible occurrences, from outages in aging infrastructure to embargos of global oil shipments to rising global demand for fossil energy. Even if we can avoid a situation in which all of these conditions happen at once, a single blackout or a period of high energy prices for fueling vehicles or for heating homes can have a vast array of negative social and economic effects. Energy planning has developed to anticipate future challenges and put a framework in place to shield society against the worst consequences of interrupted or insufficient energy supply.

As the scientific community has come to understand the causes of climate change, energy planning has become a tool to wean society from the carbon-based fuels that contribute to global climate change. Incorporating climate change goals into energy planning seems obvious, but it adds a level of complexity that fundamentally redefines the task of energy planning as originally envisioned in the late 1970s. At that time, energy planning developed in response to the oil supply crises, with less emphasis on environmental goals than today. Planning to ensure energy supplies within the existing infrastructure and concept of the energy economy is one thing; it is another task entirely to plan a vision for the future in which we have drastically reduced our reliance on this set of resources while still maintaining a high standard of living and economic production (Andrews, 2004).

This chapter explains the use of energy master planning as a tool to address anthropogenic carbon emissions, rising energy prices, and economic development. Our case study is the New Jersey Energy Master Plan, which was initiated under the leadership of Governor Jon S. Corzine and the New Jersey Board of Public Utilities (BPU) in October 2006. The authors were involved in the modeling effort in support of the plan. New Jersey has been required since 1977 to develop and maintain an energy master plan that will steer the energy supply and demand in the state toward a more sustainable, reliable, and economically sound position. The New Jersey Energy Master Plan is developed by a committee of executive branch agency and department heads and is chaired by the BPU. The current New Jersey Energy Master Plan is the result of consistent executive leadership, a broadly inclusive planning process, and extensive scenario modeling aimed at addressing issues raised by stakeholders.

Before we discuss these three elements of the current New Jersey Energy Master Plan, we explain the purpose, benefits, and challenges of shaping the future energy economy through energy master planning.

B. Purpose of an Energy Master Plan

The purpose of energy planning is to add a modicum of rationality to resources and markets that exist at a wide variety of scales and in multiple regions with vastly different levels of regulation, subsidy, and other political intervention. To complicate matters further, energy demand—energy use—results largely from individual and organizational choices rather than from a few centralized industries or locations. In short, energy planners begin their task at the confluence of a broad set of supply markets with various and sometime chaotic conditions with energy demand dispersed literally across every single person, business, and organization. This is particularly challenging when those planners are at the state or local level.

Petroleum is the resource furthest away from the control of local energy planners: about 58 percent of oil consumed in the United States was imported in 2007 (Energy Information Administration, 2009). Supplies of resources that are slightly closer, such as natural gas and coal, are also not regulated by local policymakers. Traditionally, local electricity and natural gas *delivery* do fall under the purview of local regulators; in some states, electricity generation is also regulated locally. Over the last thirty years, clean energy technologies, including both renewable energy resources and energy-efficiency technologies, have provided additional options for planners to consider. Unlike traditional energy resources, clean energy in the United States, especially in the last decade, has been encouraged and advanced primarily by state and local policymakers.

The role of the energy planner is to assess energy demand, understand the supply that currently meets that demand, consider the likely trends in supply and demand moving into the future, and, finally, influence how that supply and demand picture will develop. The purposes of the planning exercise can be manifold, though at its origins energy planning aims at sheltering the population and the economy from the negative effects of unanticipated future shocks.

In the 1970s, when energy planning first developed, these negative effects were largely economic. Environmental concerns now match the enduring economic concerns as the consequence of climate change become more and more evident. States are challenged by the prospect of rising prices for traditional energy resources as worldwide demand skyrockets and by the havoc that climate change might wreak on geography and weather patterns. Whole communities, industries, and landscapes may be in peril as a result of warming temperatures, rising sea levels, and erratic weather patterns. Examples of possible outcomes of climate change that could severely cripple local communities include Gulf Coast hurricane activity, relocating villages in Alaska's Arctic Circle, and extreme droughts and floods. Even though climate change is truly a global problem, many states are pushing forward with aggressive greenhouse gas reduction actions to remain ahead of a trend that suggests national action in the near future

The overarching goals of most energy master plans fall into general economic, reliability, and environmental categories, but the specific objectives often reflect the unique conditions of that state or region. For example, a state rich in natural energy resources may plan to continue exploiting or to conserve those resources. A state particularly vulnerable to rising sea levels may seek to put greenhouse gas emissions reductions at the center of its plan. A state with an energy-related industry—mining, drilling, or automobile production—may develop a plan that protects the jobs and economic contribution of these industries.

C. Elements of an Energy Master Plan

The art of planning lies in the details. But the goals are much larger than the details; they reflect societal mores, values, cultural trends, ideas about the past, and perspectives on the future. The role of an energy master plan, in respect to achieving goals that are much larger than the details, is to provide a comprehensive framework for policymakers on the ground. To create a plan is to take the view from above and make sure the puzzle pieces are put in their appropriate places. While the result may not be a perfect outcome, taking the time to think through the actions of yesterday, today, and tomorrow in the context of a larger policy goal is a valuable exercise in its own right. Using the goals, objectives, and strategies as a guidepost in future policymaking is the practical application of the plan. The first task in developing an energy master plan, therefore, is to define the overarching goals.

The specific energy master plan can take a multitude of formats depending on a set of key parameters and how they are construed in the plan. These parameters depend in turn on the definition of the goals. The goals can go a long distance in defining a rational time horizon for the plan, sectors of the economy to concentrate on, and stakeholder communities that should be involved.

General goals can be set without significant prior analysis. Environmental protection, increased energy efficiency, or energy reliability can serve as general goals, but giving a real framework to an energy master plan requires a *specific* goal. Determining a specific goal is aided first by understanding the historical trends and baseline conditions of the state or locality: each sector's energy consumption; the generation fuel mix; the level of imports (if any) of electricity, natural gas, or coal; the environmental impacts of energy use; and the socioeconomic and demographic trends in the state. All of these indicators help establish a structure for determining a realistic and/or aggressive goal.

Establishing a reasonable baseline forecast for the future is equally important in laying the groundwork for an energy master plan. The baseline usually assumes that trends observed historically will continue into the future, unless there is evidence to suggest otherwise. One-time events, such as the closure of a manufacturing plant or a disruptive weather event cannot be assumed to affect economic performance moving into the distant future. However, consistent growth trends in, for instance, population or vehicle miles traveled may reasonably be assumed to continue into the future.

Once the baseline is understood both for the current energy landscape and the expected future, specific goals can be articulated. Policymakers may articulate the goals unilaterally or develop them through a stakeholder process. Both methods have advantages and disadvantages; these are addressed in Section II. Regardless of the method used to determine them, the goals should reflect the chosen time horizon and take into account the implications of the baseline. This point is not made to dissuade policymakers from framing aggressive, forward-thinking goals that may have dramatic consequences for the energy economy of their states. Rather, a plan that sets out goals that are technologically or economically infeasible within a given time horizon or that ignores natural resource limitations or existing industries within a state is less likely to be achieved or to gain traction with important stakeholder communities.

Articulating goals alone cannot create a functional energy master plan. Goals need objectives that are specific and direct. In our case study, the New Jersey BPU approached the planning process and the involvement of stakeholders through the lens of objectives and strategies with responsible parties, known costs, expected milestones, and defined timelines for implementation. Though it will take a number of years to validate whether this approach has paid off for New Jersey, the rationale is clear for providing as much specificity as possible from the outset.

A completely rational planning process still faces challenges, the first among these being the impossibility of predicting the future. Trends in energy prices may suddenly reverse themselves, as has happened in the past year, demand for a certain energy commodity may explode globally, or the leadership for or public perception of the need for energy planning may evaporate. Moreover, plans with goals and objectives require significant implementation strategies and future action. If this future action does not materialize, the plan becomes a vision of the future once articulated and then ignored.

An inappropriate response to uncertainty, however, is to assume that planning has no value. The value of planning is in establishing the conditions under which policies will be successful and thereby in defining the framework that would need to be adjusted if those conditions fail to materialize (Center for Energy, Economic & Environmental Policy, 2008).

D. Case Study: New Jersey Energy Master Plan

Before developing the specifics of the New Jersey Energy Master Plan a variety other state's energy master plans were reviewed. Many U.S. states have undertaken the task of developing and implementing energy master plans. Among the states with active plans are California, New York, Vermont, West Virginia, and Florida. Different states take different approaches to energy planning.

- ***California*** – California's 2003 "Integrated Energy Policy Report" and statutory framework relate most directly to system and grid reliability, with a secondary but strong focus on conservation and environmental stewardship. As one of the ten largest economies in the world, California relies heavily on reliable energy production, distribution, and cost. The main energy planning law, passed in 2002, responds directly to the state's experience with energy market deregulation that culminated in 2001 with spiking prices and rolling blackouts during some of the hottest summer weeks (California Energy Commission, 2003).
- ***New York*** – The New York State Energy Research and Development Authority published its "2002 State Energy Plan and Final Environmental Impact Statement" with the goal of keeping the state in an economically competitive position through

investments in new and emerging energy technologies (New York State Energy Research and Development Authority, 2002).

- ***Vermont*** – Published in multiple volumes beginning in 1998, Vermont's "Fueling Vermont's Future: Comprehensive Energy Plan and Greenhouse Gas Action Plan" focuses aggressively on addressing greenhouse gas emissions through energy efficiency, conservation, and renewable energy goals (Vermont Department of Public Service, 1998).
- ***West Virginia*** – In a state with significant fossil fuel resources, West Virginia's plan seeks to cut the state's dependence on imported oil and protect the state's economy through more innovative uses of coal resources, promotion of integrated fossil and renewable energy technologies, and promotion of energy efficiency (West Virginia Division of Energy, 2008).
- ***Florida*** –This state's energy master plan was developed partially in response to the extreme hurricane events of 2004 and 2005. The plan emphasizes diversifying fuel supply through a series of voluntary actions supported through tax incentives (Florida Department of Environmental Protection, 2006).

New Jersey joined this group in October 2006 with the commencement of the energy master planning process by Governor Jon S. Corzine. New Jersey's Energy Master Plan has been a statutory obligation of the State since 1977 when the legislature passed the law (52:27F-14), that created the New Jersey Department of Energy and the mandate to write energy master plans to direct energy supply and demand issues in the state. The law was amended in 1987 to create a standing Energy Master Plan Committee and define its membership and their roles. This committee was to be comprised of commissioners or representatives from several State agencies, including Commerce, Energy and Economic Development; Community Affairs; Environmental Protection; Health; Human Services; Transportation; and Treasury.[1] The task of the committee as stated in the 1987 amendment to the law was to:

> "... prepare or cause to be prepared, and after public hearings as hereinafter provided, adopt a master plan for a period of 10 years on the production, distribution, consumption, and conservation of energy in this State. Such plan shall be revised and updated at least once every three years. The plan shall include long-term objectives but shall provide for the interim implementation of measures consistent with said objectives (NJ State Law, 1987)."

The law also laid out the requirements for public participation in the process and the role of the committee in ensuring that the State and affected agencies comply with the goals of the master plan. However, the law did not include actual mechanisms for the committee or the participating agencies to force compliance by the State or each other, thus necessitating follow-up legislative or regulatory action.

The essence of the law was to create a far-reaching and comprehensive document to serve as a guidepost across several State agencies for a period of at least ten years. The three major

1 The Department of Commerce, Energy, and Economic Development no longer exists. Economic development issues are handled by the Economic Development Authority. Energy is handled by the Board of Public Utilities or the Department of Environmental Protection, depending on the issue.

goals defined in the 1977 statute are energy security, economic growth, and environmental protection (NJ BPU, 2008). The plan should reflect not only the wishes and perspectives of the committee members and their agencies, but also the thoughts of the general public captured through public hearings. Incorporating regular updates into the law recognizes the dynamism of the market and planning context and the need to revisit the plan on a periodic basis.

The most recent comprehensive New Jersey Energy Master Plan before the current efforts was released in 1991.[2] It compiled then-current data on energy use in the state and made general statements about directions that could be taken as well as seventy-seven specific recommendations (New Jersey Department of Environmental Protection and Energy, 1991). [3] These recommendations were intended to work toward four broad goals: (1) "To provide secure energy supplies and services to energy users"; (2) "To encourage economic growth by providing energy services at the least cost"; (3) "To protect the environment through wise and efficient energy use"; and (4) "To balance energy needs and impacts through coordinated and integrated planning." (New Jersey Department of Environmental Protection and Energy, 1991)

The process of creating the 1991 New Jersey Energy Master Plan included extensive participation by the public and interested stakeholder organizations. One major public hearing included more than 500 participants representing more than 80 groups and individuals (New Jersey Department of Environmental Protection and Energy, 1991). The published plan included their inputs and also included general recommendations, but was somewhat vague on presenting exact recommendations for implementation. This was addressed with the second part of the report published in 1992 as the "New Jersey Energy Master Plan Implementation Section". This smaller document distilled the seventy-seven recommendations into concrete policy recommendations. For example, one recommendation read: "The Department of Transportation should facilitate greater use of vanpooling and carpooling." The Implementation Section expanded on this recommendation by adding three specific suggestions: continue and expand the existing ridesharing program; assist local and suburban transportation management associations in developing van- and carpooling programs in their areas; and review the possible need for additional "park-and-ride" lots to increase the convenience of car- and vanpooling programs.

Overall, the process and outcome of the 1991/1992 composite New Jersey Energy Master Plan was in line with the strategic and communicative intention of the law. The four broad goals were then separated further into logical focus areas (i.e., the goal of reducing energy dependency is quantified through a further division to "efficiency improvements," "renewables," etc.) and goals within these areas were identified. The Implementation Section distilled these recommendations into manageable pieces. Public participation was central to the process:

> "The Energy Master Plan Committee will continue to enlist the kind of broad-based public participation that is crucial to the successful implementation of public policy goals. As each member-department [of the Committee] begins to implement Plan policies

---

2 There was a draft report in 1981, which was never adopted as state administrations transitioned. There were also 1979 and 1985 Energy Master Plans which were unilaterally adopted by the NJ Department of Energy.

3 These are presented as "findings." For example: "The state can mitigate its vulnerability to supply disruptions and cost escalation by reducing its high dependence on petroleum products."

that have direct bearing that department's statutory functions and responsibilities, the public will again be invited to participate in the rulemaking process (New Jersey Department of Environmental Protection and Energy, 1991)."

The 1995 New Jersey Energy Master Plan has a distinctly different interpretation of the law and the role of the State in directing the energy markets toward meeting the goals of the master plan. The 1995 "New Jersey Energy Master Plan: Phase I Report" reflects two major changes in the state politically and in the energy markets. First, the inauguration of a more market-focused, Republican administration shifted the stance of the Office of the Governor and State agencies away from heavy intervention or regulation toward a more *laissez faire* approach. Second, imminent deregulation of the energy markets and the entrance of multiple energy providers may have lowered the capacity of the State to exercise certain kinds of oversight that had been available when utilities were primarily regulated monopolies. Thus, the 1995 New Jersey Energy Master Plan is a very different document than its predecessor.

This report was intended to be the first of several phases in the creation of an overall energy master plan for New Jersey. In contrast to the 1991 plan, however, it is not a large document that provides a complete picture of demand and supply for energy in the state. Instead, it is a smaller step in the direction of first setting goals and suggesting some general directions to take. Above all, it emphasizes the diminished role of the State government in regulating energy markets to achieve the goals of the plan. This diminished role was described as due mostly to several pieces of federal legislation and action that were increasingly defining the energy landscape nationally and in the state, and subsequent state reactions.[4]

The Phase I Report explains the federal legislation and suggests how to "manage the transition to competition (New Jersey Board of Public Utilities, 1995)." Included in this report are some specific recommendations. One of the main recommendations is the adoption of integrated resource planning (IRP) techniques by the individual utilities, with the aim of streamlining the state regulatory arm and shifting the onus for increased efficiency to the utility level.

After 1995, New Jersey did not follow through on implementation or updates of any suggested policies of the New Jersey Energy Master Plan. The election of Governor Jon S. Corzine in 2005 and the leadership within the Board of Public Utilities, though, have changed this situation. Governor Corzine ran his campaign on the promise of a more comprehensive energy policy, including aggressive energy efficiency goals, and the promise of addressing greenhouse gas emissions in the state. This appealed directly to the fact that energy has been on the minds of residents and business owners in New Jersey for several years. The events of September 11, 2001, which disproportionately affected residents of New Jersey and New York, brought the nation's reliance on foreign oil into focus. Two years later, the East Coast blackout of August 2003 revealed the fragility of the electric generation and transmission system. Natural gas prices began to increase in the early 2000s, showing little sign of easing. New Jersey has relied increasingly on imported electricity, which affects the average price

4 These include, for example, Orders 436 and 636 of the Federal Energy Regulatory Commission in 1985 opening customer access to natural gas markets; the Public Utilities Regulatory Policies Act of 1978; the Energy Policy Act of 1992; the Clean Air Act Amendments of 1990; and federal deregulation of the telecommunications industry.

that consumers face. All these issues underscore the need to think proactively and aggressively about developing a new energy economy.

It was under these circumstances that the 2008 New Jersey Energy Master Planning process began. As is discussed further below, the 2008 New Jersey Energy Master Plan had to be developed in the context of regional wholesale electricity markets. In addition, the plan extends past market-like policies such as a carbon dioxide cap-and-trade system and tradable renewable energy credits as part of a state renewable portfolio standard. Thus, the differing approaches in the 1995 New Jersey Energy Master Plan and the 1991 Plan are brought together in the 2008 Plan.

II. Leadership and Process in the New Jersey Energy Master Plan

The 2008 New Jersey Energy Master Plan is the result of significant executive leadership and extensive stakeholder and public involvement. The spirit of the New Jersey Energy Master Plan law is to create a plan that changes energy consumption behavior in the state by setting the framework for regulators and legislators to develop a comprehensive set of policies that will achieve the plan's goals. In the past, plans have been promulgated, but the follow-through actions have not taken place. Part of this can be attributed to the level of executive interest in the development of the plan. Though every plan thus far released has had the official seal of the Governor's Office on it, the impetus for them really came from the Board of Public Utilities and its precursor agencies. In addition, broad public and stakeholder participation in the plans were not placed front and center. Aside from the requisite public hearings, the previous energy master plans were not, by and large, the result of an inclusive process.

A. New Jersey's History of Environmental Leadership

New Jersey has a significant history of progressive environmental leadership, especially in the last twenty-five years. Here are a few examples (Andrews, 2006).

- *Statewide Source Separation and Recycling Act (1987)* – The Act required all counties to develop a comprehensive recycling plan aimed at reducing municipal solid waste by 15 percent in the first year and 25 percent in subsequent years. This law was amended in 1992 to increase the goal to 60 percent by the end of 1995. In addition, a temporary tax of $1.50 per ton of solid waste deposited in landfills was established. The funds were deposited in the State Recycling Fund and allocated for a variety of special projects including: municipal and county recycling tonnage grants; low interest loans or loan guarantees to recycling businesses and industries and recycling market development research; public information and education; county recycling program grants; and state recycling program planning.
- *Renewable Portfolio Standard (1999)* – New Jersey adopted a renewable portfolio standard as part of its electric industry restructuring law. The current requirement,

which was adopted in 2006, requires 22.5 percent of electricity sold in the state in 2020 (energy year 2021) to come from renewable energy resources. There is a separate solar energy carve out of 2.12 percent and a cap on "Class II" renewable resources of 2.5 percent. As part of the program, renewable energy credits (RECs), and, in a separate market, Solar credits (S-RECs), are traded to offset the higher cost of these technologies. In addition to the REC markets, the New Jersey Clean Energy Program was established to offer a variety of incentives to offset the initial capital investment for the installation of clean energy technologies. The program is funded through the State's Societal Benefits Charge. Together, these incentives have placed New Jersey second only to California in supporting the solar energy industry.

- *State Development and Redevelopment Plan (2001)* – Since the 1985 passage of the State Planning Act (N.J.S.A. 52:18A-196 et seq.), natural resource preservation and urban development and redevelopment issues have been guided in part by the State Planning Commission. The most recent Plan, promulgated in 2001, designates smart growth development areas and protected areas, with a range of categories in between. Growth is encouraged in smart growth areas through a set of incentives and discouraged in protected areas with special fees and other disincentives. The goals are to promote high- and mid-density development with greater access to transit and to dissuade greenfield development.
- *Office of Smart Growth* – The Office of Smart Growth (OSG) coordinates planning in New Jersey to protect the environment and guide growth toward compact, mixed-use development and redevelopment. The Office implements the goals of the State Development and Redevelopment Plan through comprehensive, long-term planning. It helps realize the goals of the plan by working toward programmatic and regulatory land-use decisions at all levels of government and the private sector.
- *Transit Villages Initiative* – This initiative is a multi-agency collaboration headed by the New Jersey Department of Transportation and NJ Transit. It aims to promote mixed-use, dense development around existing transit hubs to increase transit ridership, decrease traffic congestion, and promote environmental stewardship. A community must meet a number of criteria to be designated as a transit village. With transit village designation, municipalities are eligible for annual grants to improve their communities adjacent to the transit hub. There are currently nineteen designated transit villages.[5]
- *Line Extension Rules (2002)* – As part of the State's Development and Redevelopment Plan, State agencies were charged under Executive Order No. 2 of the McGreevey Administration to implement rules that are in keeping with the State's smart growth goals. To this end, the Board of Public Utilities passed the line extension rules that require developers who choose to develop outside of designated smart growth areas to bear the costs of extending utility infrastructure.

---

5 New Jersey Transit Villages include Belmar, Bloomfield, Bound Brook, Burlington, Collinswood, Cranford, Elizabeth, Jersey City (Journal Square), Matawan, Metuchen, Morristown, Netcong, New Brunswick, Pleasantville, Rahway, Riverside, Rutherford, South Amboy, and South Orange. The majority of these communities are on major train lines either to New York City or Philadelphia.

B. Role of the Executive in Framing New Jersey's Energy Master Plan

From the beginning of the current plan's process, the Governor's Office directed the effort, publicized the goals, and encouraged active participation by the public and stakeholder groups. The Board of Public Utilities has chaired the committee and hosted several stakeholder meetings and elicited comments through public forums and online comment forms. In this section, we explain the process and comment on the strengths and weaknesses of the approach.

The overarching goal of the New Jersey Energy Master Plan resulted from one of the Governor's campaign promises, namely that he would ensure a reduction in New Jersey's fossil energy consumption 20 percent below the forecasted amount in year 2020 through the use of energy efficiency. This was perhaps the most ambitious energy use goal of any state at the time.

While the BPU was responsible for most of the organizational and process framing tasks, the Governor's Office of Economic Growth was the main executive driver of the plan. The Governor's leadership was expressed through the definition of the overarching goal (20 percent reduction in projected energy demand) and by communicating the main economic concerns of the administration: New Jersey's energy prices relative to those in the rest of the region, and the impact of energy prices on job creation.

In the early months of 2007, the New Jersey Energy Master Plan was further defined by the Governor's Executive Order 54, which laid out aggressive greenhouse gas reduction goals for 2020 and 2050 (Executive Order 54, 2007). These goals were reiterated in the Global Warming Response Act passed by the legislature and signed into law by Governor Corzine in the summer of 2007 (New Jersey Assembly Bill No. 3301, 2007). The Act requires the state to reduce greenhouse gas emissions to 1990 levels by 2020 (roughly a 20 percent decrease) and to 80 percent below 2006 levels by 2050. These goals reflect the United Nation's Intergovernmental Panel on Climate Change (IPCC) consensus on the reductions necessary to stabilize the climate (Intergovernmental Panel on Climate Change, 2007). The BPU, in concert with the committee agencies and stakeholders, developed a comprehensive set of objectives and strategies aimed at reducing energy use and greenhouse gas emissions by 20 percent by 2020.

The proposed strategies can be divided into three main types. First are energy efficiency strategies designed to improve the performance of appliances and buildings through the use of technology and smarter, more energy efficient building practices. Specifically, the plan suggests that the State should pass advanced appliance standards, regulate the energy efficiency level of new construction, implement demand-response programs with commercial and industrial customers, and introduce a widespread, utility-based program aimed at improving the efficiency of existing buildings through comprehensive "whole building" efficiency improvements. Second, the plan aims to reach the goals by replacing some conventional energy supplies with renewable and alternative generation resources such as combined-heat-and-power (CHP). Several pilot projects in solar, on-shore wind, off-shore wind, biomass, and behind-the-meter CHP are intended to reduce carbon emissions and create jobs. Finally, the plan considers conventional energy supply and how to create a more sustainable portfolio of conventional energy resources.

C. Executive Agencies

The Energy Master Plan Committee was comprised of a set of key State agencies particularly related to energy, the environment, transportation, and the health and welfare of citizens. The Committee was chaired by the BPU, but it was to develop the plan cooperatively by bringing the expertise and perspective of each agency to the table. The most active agencies in the development of the 2008 plan were the BPU, the Department of Environmental Protection, the Department of Community Affairs, the Department of Transportation, the Department of Labor and Workforce Development, and the Department of the Treasury. The involvement of these departments, unlike that of the BPU, had more to do with directives given to the agencies individually by the Governor's Office and less to do with the legal framework of the New Jersey Energy Master Plan. The Department of Environmental Protection was given the task of putting together a greenhouse gas inventory required to set the framework for the implementation of the Global Warming Response Act. Treasury was asked to assess the State government's own energy consumption behavior and look for ways to trim its demand. The Department of Community Affairs was generally in charge of adopting and revising international and national building code standards; it became involved in the planning process once building codes were suggested as a possible strategy. The Department of Labor and Workforce Development was charged with assessing the training program infrastructure in the state and developing recommendations for bolstering educational programs that specifically support new, "green" industries that will develop, it is hoped, in response to the New Jersey Energy Master Plan policies.

D. Stakeholder Involvement

About six weeks after the Governor announced the start of the New Jersey Energy Master Plan process, several working group meetings were convened by the Board of Public Utilities. These working groups focused on renewable energy, energy efficiency in electricity and space heating applications, electricity imports, and peak demand. As regulators of electric and natural gas utilities, the BPU remained focused on these areas, allowing the Department of Transportation and the Department of Environmental Protection to take the lead on transportation and environmental issues, respectively. Approximately 500 stakeholders attended these meetings. They represented utilities, non-profit groups, environmental organizations, and organizations of concerned citizens. The stakeholders were organized into working groups and given about a month to put together reports for the BPU. These reports were to contain their comments on the baseline from their subject matter perspective, thoughts concerning future trends, and recommendations for which issues were the most pressing to be addressed in the plan and how that State should approach them.[6] The reports were used to frame an initial set of policy strategies.

After these initial stakeholder meetings, stakeholder participation became more targeted. On a consistent basis, utility representatives, environmental groups, and small businesses involved in renewable energy technologies took part in aspects of forming the plan. The

6 All working group reports can be accessed by visiting the official New Jersey Energy Master Plan website at www.nj.gov/emp

motivation for expanding the involvement of stakeholders was first to improve the quality of the plan with expert input and second to ensure a certain amount of buy-in for the final plan at the time of its release. As a broad-reaching plan, all parts of society will be called on to be involved in its implementation, either directly or indirectly, and therefore widespread support for the plan was from the beginning viewed as of paramount importance. By and large, stakeholders applauded the effort to create an energy master plan and being asked to be involved in the process. Many groups were never able to agree completely with the proposed policies, but nevertheless supported the overall effort.

In April 2008, a draft of the plan was released and thus subjected to public scrutiny, including through official public hearings. Several issues were raised during this period. First, the amount of renewable energy included in the plan was questioned heavily by the businesses involved in developing these technologies in the state. This group argued that using the strict percentage guidelines of the Renewable Portfolio Standard Board Order would serve to shrink the overall amount of renewables in the New Jersey Energy Master Plan as a result of the declining sales of electricity. If a certain capacity of renewable resources was deemed possible in the Business as Usual (BAU) scenario, they argued, this capacity should be at the very least maintained under the New Jersey Energy Master Plan. Second, the draft plan suggested the possibility of creating a new State organization that would be in charge of managing the energy efficiency programs. Critics cited multiple examples of beleaguered State bureaucracies and general dissatisfaction with the idea of creating a new agency in a state where misuse of public funds has, at times, been rife. Third, conventional electricity generation was considered in the draft plan in a way that suggested the possibility of a new or upgraded nuclear facility in New Jersey. Environmental and concerned citizens groups criticized this as a short-sighted view that did not consider the dangers associated with nuclear power and nuclear waste. Finally, business groups and the Public Advocate expressed concerns over higher retail energy rates that could have a negative impact on industry and low-income households in the state. Many of these concerns were directly addressed in part or in whole in the final plan, which was released with significant positive reaction in October 2008.

E. Challenges in Process and Implementation

The process of developing the New Jersey Energy Master Plan was successful for two main reasons. First, the executive leadership was sustained and clear from the beginning. Without this direction and prioritization, the plan might have faded from public view and developed very differently, if at all. Second, stakeholder participation was recruited and maintained throughout a process that required double the time than was originally planned. Yet, despite these successes, the plan will likely face significant challenges as it moves into the implementation stage.

Along with the New Jersey Energy Master Plan, an organized compendium of strategies for achieving the objectives was released. The strategies identify the future actions needed for the successful implementation of the policies. Some actions are regulatory and some are legislative. All require efforts beyond the tenure of current leadership, and thus depend, to a certain extent, on the buy-in of future generations of regulators and legislators. One antidote to this nontrivial problem could be consistent executive leadership. Creating the New Jersey

Energy Master Plan was a full-time task for a number of BPU staff members for the duration of the project. Implementation will take at least these resources spread across the various parts of the government and the private sector. A person or group entrusted with general oversight of all the activities could improve the likelihood of the activities' success. Likewise, consistent oversight could help to catch strategies that are not meeting expectations and work to redefine them.

Future generations of State leaders and policymakers may also face unanticipated, game-changing conditions that make some of the basic tenets of the plan irrelevant or unworkable. The plan, though designed to be revisited on a triennial basis, will require committed leadership and resources to undertake its ongoing reconsideration and revision. A dedicated officer or office assigned to New Jersey Energy Master Plan oversight would arguably be more able to react and accomplish a significant redrafting of the plan than would be achieved by attempting to reconvene the diverse group of civil servants involved in the development of the current plan.

III. ECONOMIC MODELING TO SUPPORT THE NEW JERSEY ENERGY MASTER PLANNING EFFORT[7]

An extensive quantitative modeling process accompanied the development of the 2008 New Jersey Energy Master Plan. Models, as the planning process itself, are by no means perfect, but nonetheless can add a level of analytical legitimacy and address concerns of critics. The modeling effort was conducted with significant involvement from stakeholders and in a largely transparent manner.

The Center for Energy, Economic and Environmental Policy (CEEEP) and the Rutgers Economic Advisory Service (R/ECON™), both located within the Edward J. Bloustein School of Planning and Public Policy of Rutgers, the State University of New Jersey, were commissioned by the BPU to provide data and modeling support for the master plan effort. This section describes the context, data assumptions, and final calculations made by CEEEP and R/ECON™. The goals of the models were to capture the relative macroeconomic impacts of the New Jersey Energy Master Plan compared to the Business as Usual case. The work conducted by these two organizations was preceded by a New Jersey State energy targeting project involving regulators, stakeholders, and researchers, including some from R/ECON™ and CEEEP (Andrews, Jonas, Mantell, and Solomon, 2008).

A series of prior events helped build the foundation for the modeling. On December 18, 2006, CEEEP and R/ECON™ presented the modeling framework used in the master plan modeling effort to stakeholders. On January 5 and 19, 2007, CEEEP convened two technical working groups to elicit input on electric generation and transmission. In addition, CEEEP and R/ECON™ participated extensively in many stakeholder meetings convened as part of the New Jersey Energy Master Plan process from late 2006 through September 2008.

Section A below discusses the role of modeling in long-term energy planning; Section B articulates the energy modeling used in the master planning process; Section C describes the two major models that were used in the process; and Section D presents the results.

7 Much of this section is based upon the Center for Energy, Economic and Environmental Policy's "Modeling Report for the New Jersey Energy Master Plan".

A. Long-Term Energy Planning and Modeling

The value of planning comes from both the process and the outcomes. The process provides a structure that should help policymakers and stakeholders think through implications and impacts of different strategies. The output from modeling the plan enables policymakers and stakeholders to assess whether the plan satisfies its objectives, what its likely impacts on the economy and the environment will be, and how it will perform over time.

A cursory review of energy events over the last several decades reveals that the unexpected is the norm, not the exception. In the late 1970s and early 1980s, there were serious concerns about the possibility of oil prices reaching $100 per barrel. During the 1990s, oil and natural gas prices were at very low levels, but they have increased dramatically in recent years. In the 1970s, natural gas was not permitted to be used to generate electricity. In the 1990s and until the relatively recent spike in natural gas prices beginning in 2002, natural gas became the dominant fuel for new generation plants. In 1979, the meltdown at Three Mile Island precipitated a halt in the construction of new nuclear power plants; now, a possible resurgence in the use of nuclear power may be occurring, with the extension of licenses by twenty years and preliminary plans for building new plants. Air emissions concerns in the 1970s revolved around sulfur dioxide and nitrogen oxide; today, they also include emissions of greenhouse gases like carbon dioxide and methane. Finally, new technologies are being developed and improved, including those for hybrid vehicles, fuel cells, carbon sequestration, biomass, wind turbines, and solar power.

The New Jersey Energy Master Plan assumes a critical connection between energy, the environment, and the economy; it assumes that changes in all three must be studied and addressed during the planning process. In addition, the plan assumes that fossil fuel prices are likely to increase over the long term, global warming is a serious problem that requires immediate action, energy efficiency is the most cost-effective means of addressing most of the foreseeable future increases in energy needs, and a variety of renewable technologies should be developed and implemented for economic and environmental reasons. It is within this context that the role of modeling is discussed.[8]

B. Energy Modeling in Support of the New Jersey Energy Master Plan Process

This section reviews the modeling effort that supports the development of the New Jersey Energy Master Plan. It discusses the purpose and limitations of the modeling effort, the distinction between the modeling efforts and the New Jersey Energy Master Plan, the difference between planning and forecasting as it relates to the New Jersey Energy Master Plan, and the relationship between modeling results and implementation.

The purpose of modeling is to inform the process, not to be dispositive. The engineering, economic, and policy issues are so complex and intertwined that there is no single "correct" solution that the modeling can calculate. Rather, by comparing alternatives, the simulations using the models illustrate the likely differences in outcomes under different sets of

8 As of early December 2008, oil prices were less than $50 a barrel due to the global economic slowdown.

defensible, reasonable assumptions. It forces data collection and analysis, justification of assumptions, understanding of complexities and relationships, and rigorous ways to test intuition and establish orders of magnitude. Regardless of the results, the modeling process is extremely helpful. The modeling may narrow areas of disagreement, help identify uncertainties that matter and those that do not given the policy choices, identify key tradeoffs, and establish the conditions under which certain outcomes can occur. The intent is that the results themselves contribute to the planning process and discussions, not that the model determines the specific policy design of the final plan.

The New Jersey Energy Master Plan should not be confused with the assumptions and policies modeled. Clearly, there is an important connection between a proposed plan and the modeling effort, but that does not mean that what is modeled is "the Plan." For example, the modeling effort may assume that 3,000 mega-Watts (MW) of New Jersey off-shore wind are built by the year 2020, but the Plan, as implemented, may begin with a smaller pilot facility to gain more knowledge and experience about costs and performance. The results from the pilot project will have implications regarding whether and how to continue.

Furthermore, the modeling cannot reproduce every possible policy or investment alternative because there simply is not enough time or analytical resources to do so. Moreover, the models may not be sensitive to the differences between two *similar* policies that have significant implications for one stakeholder group versus another. Understandably, stakeholders are concerned about particular projects, and may infer that the inclusion or lack of inclusion of a particular project in the modeling indicates that the New Jersey Energy Master Plan does or does not consider that project as part of the Plan.

In addition, many stakeholders have strong views about what the future holds. Naturally, they would like to see these views embodied in the New Jersey Energy Master Plan and modeling assumptions as much as possible. For instance, many stakeholders believe that a national carbon dioxide policy will be implemented during the planning horizon of the New Jersey Energy Master Plan. This belief has been justified by legislation that has moved through the federal government. Because of this, the modeling effort is based on existing or explicit state- or regionally driven policies—in this case, the Regional Greenhouse Gas Initiative (RGGI), as well as impending national legislation. For the purposes of New Jersey Energy Master Plan modeling, RGGI is utilized as the carbon dioxide policy for the years 2010 and 2015 and a national carbon dioxide policy is used for the year 2020.

Another frequent source of misunderstanding is equating modeling with forecasting. The modeling effort compares the BAU and Energy Master Plan scenarios under two reasonable sets of assumptions. The BAU and Energy Master Plan scenarios are meant to represent four possible energy futures for New Jersey. The BAU follows the current trajectory that New Jersey established in the 2004 base year, and the BAU Pessimistic follows that same established policy trajectory under higher (worse) national and international energy price conditions. The Energy Master Plan scenarios represent a set of energy futures that depart significantly from the current path. The purpose of the modeling effort is to provide quantitative calculations that inform decisions regarding the comparison of these two scenarios by helping to determine which policies have a relatively greater impact on the state's energy, environmental, and economic landscape than others.[9]

9 This is a typical approach in policy analysis. See, for example, MIT Study on the Future of Coal, 2007, pp. 8-9, available at http://web.mit.edu/coal/.

This is not meant to suggest that in developing the BAU and Energy Master Plan scenarios assumptions are chosen without care. A great effort has been made to pick reasonable, credible, and objective assumptions for use in the models. That being said, since the purpose is to compare the relative merits of two courses of action, it is the relative differences that drive the comparison, not absolute numbers. In contrast to this relative analysis, investors in projects and other stakeholders care significantly about absolute outcomes, such as the price of natural gas or the capacity factor of a wind farm.

Another common misunderstanding is equating each assumption to an explicit provision of the plan. In many cases, the models require a level of detail well beyond what is appropriate for a long-term plan. For instance, in modeling future electricity prices, the model that is used calculates the amount of electricity a power plant produces in every hour in the year 2020. Clearly, one should not interpret the output of this model to mean that New Jersey wants power plant X to produce 123 MW at 1 pm on July 23, 2020.

In other cases, a sufficient level of detail regarding a particular issue is not readily available, and so assumptions must be made. An example of this is the judicious location of energy efficiency measures, demand response, and combined heat and power facilities that may allow the postponement or avoidance of transmission and distribution (T&D) upgrades. Determining these locations requires highly specific information that is not readily available. Given this practical limitation, the modeling made generic assumptions about the potential T&D savings that could occur with judicious location of these measures.

It is also important to consider the modeling results in the context of the underlying strategies. Sometimes the relationship between policies and outcomes can get lost in all the calculations. The modeling provides a means to test and understand the Plan's themes and strategies, not become the focus of the Plan.

The modeling process establishes many of the conditions under which policies can achieve their intended outcomes. This helps tremendously with evaluating policy implementation and understanding the conditions under which policy changes may be necessary. Some assumptions made in the modeling will be wrong. Unforeseen events will require changes in direction and policies. Planning in general and modeling in particular can help anticipate these possibilities and determine appropriate responses at the appropriate times.

In short, the New Jersey Energy Master Plan must explicitly deal with uncertainty and the prospect that things will turn out differently from what was modeled. This often gets lost in the discussions, as modeling is frequently assumed to be a forecasting effort with definite outcomes. The data and modeling assumptions have associated ranges of uncertainties. Even in situations in which one would think the range of uncertainty should be small—for example, the cost of a combustion turbine—they can be surprisingly large. These uncertainties need to be considered when evaluating calculations. Although models calculate numbers to a precise value, this "precision" is a programming artifact and must be understood as such. What also should be kept in mind is that the range of uncertainty varies with specific assumptions. The uncertainty in the cost of a combustion turbine is less than the uncertainty of the cost of off-shore wind, which is in turn less than the uncertainty associated with the cost of a new nuclear power plant.

The primary drivers for the modeling calculations are the assumptions about the cost and magnitude of energy efficiency and demand response for electricity and natural gas. If one assumes that energy efficiency and demand response are cost-effective (which numerous studies have concluded) and that state policies can influence energy efficiency and demand response successfully, then one does not need modeling to conclude that energy bills will decrease, environmental impacts will be lessened, and the New Jersey economy will not be harmed. The modeling provides the order of magnitude, confirms the intuition, and helps target policies that can help to make these outcomes more likely.

C. Description of Models

Two major models are used as part of this effort. The first is DAYZER (Day-Ahead Locational Market Clearing Prices Analyzer), a sophisticated model of the PJM Interconnection (PJM) day-ahead wholesale electricity power market of which New Jersey is a part.[10] DAYZER is a unit commitment and dispatch model that mimics PJM as closely as practical, including calculating the locational marginal prices (LMPs) that vary by location and time. The second is R/ECON™, a detailed econometric time-series model of the New Jersey economy. The results from DAYZER, along with many other assumptions, are provided to R/ECON™ as inputs, so that it can simulate the state's economic future under a variety of energy scenarios.

1. DAYZER – The PJM Wholesale Electricity Market Model

DAYZER calculates locational market clearing prices and the associated transmission congestion costs in competitive electricity markets.[11] This tool simulates the operation of the PJM electricity market—the dispatch procedures adopted and used by PJM—and replicates the calculations made by PJM in solving for the security-constrained, least-cost unit commitment and dispatch in the day-ahead markets. The LMP and congestion cost calculations are based on data on fuel prices, demand forecast, unit and transmission line outages, and emission permits costs. DAYZER incorporates all the security, reliability, economic, and engineering constraints on generation units and transmission system components.

DAYZER has the following features:

- Provides accurate security-constrained unit commitment and dispatch algorithms that mimic those used by PJM in the day-ahead market;
- Enables accurate modeling of PJM with its own particularities (second contingency constraints, locational reserve markets, etc.);
- Captures marginal transmission losses in dispatch and clearing prices;
- Captures transmission outages, transmission contingencies, nomograms, and planned and known transmission upgrades;
- Models accurately phase angle regulators and loop flows;

10 PJM Interconnection is a regional transmission organization (RTO) that coordinates the movement of wholesale electricity in all or parts of Delaware, Illinois, Indiana, Kentucky, Maryland, Michigan, New Jersey, North Carolina, Ohio, Pennsylvania, Tennessee, Virginia, West Virginia, and the District of Columbia

11 DAYZER was developed by Cambridge Energy Solutions (http://www.ces-us.com/index.html). DAYZER was used in a recently commissioned study by PJM and Mid-Atlantic Distributed Resources Initiative on estimating the benefits of demand response. See The Brattle Group, Quantifying Demand Response Benefits in PJM, January 29, 2007

- Allows users to analyze various scenarios and quantify the impact of key variables/assumptions;
- Employs random outage using Bernoulli Probability modeling;
- Enables the optimization of generation maintenance schedule based on reserves; and
- Uses import and export schedules to account for flows to and from neighboring markets

DAYZER requires that both transmission and generation additions and retirements be input exogenously into the model.[12] The existing PJM transmission system is used in the DAYZER runs.

In the current modeling effort, generation expansion plans are based on the following process: PJM's load forecasts by zone by year are used to calculate the hourly loads using PJM's 2006 load duration curve. The amount of systemwide installed capacity is calculated based on PJM's 15 percent reserve margin. Renewable generation needed to meet individual state's Renewable Portfolio Standards (RPS) is then included in the expansion plan. If additional generation is needed to meet the installed reserve margin, it is added. The type (baseload, intermediate, or peaking) and the fuel (nuclear, coal, or natural gas) are determined by reviewing the PJM generation interconnection queue for each particular PJM zone. Historically, the PJM generation queue contains more generation than is actually built. DAYZER is then run using the candidate expansion plan to ensure that generation unit capacity factors are appropriate for the type of unit and to ensure there are no hours in which demand exceeds supply in each zone that DAYZER tracks. In addition, locational marginal prices and net operating revenues are checked to ensure that either retirements or new generation would not otherwise occur. Modifications to the candidate expansion plan are made as necessary, and DAYZER is re-run until a satisfactory expansion is developed.

In developing the Energy Master Plan, four major scenarios were evaluated during the Plan's time horizon, which ends in the year 2020. The BAU scenarios represent two specific possible futures whose outcomes reflect New Jersey continuing on its current course with respect to energy policy. There are, of course, many other possible outcomes that are numerically different from the specific BAU outcomes modeled but are similar in nature.

The Energy Master Plan scenarios utilize a combination of programs designed to implement energy efficiency, renewable energy, and other actions that are intended to achieve the greenhouse gas emission limits specified in the New Jersey Global Warming Response Act and the RPS requirements for 2020.

When comparing the outcomes of the BAU and Energy Master Plan scenarios, there is an important caveat: any comparison must include the reduction in greenhouse gas emissions that is anticipated in the Energy Master Plan scenarios compared to the BAU scenarios. Comparing the economic performance of these two scenarios without accounting for the economic value of the greenhouse gas reductions has the effect of making the BAU scenarios appear economically favorable to the Energy Master Plan scenarios. The difficulty is that there is a wide range of estimates regarding the negative economic impact per CO_2 ton. The IPCC's Working Group II estimates a range from \$3 to \$95 per ton based upon a survey of 100 estimates with a mean of \$12 per metric ton (Metcalf, 2007). The Stern Report estimates

12 Some models have the ability to construct generation expansion plans but do not have the detailed locational marginal price capabilities of DAYZER.

the social cost of CO_2 at $85 per ton (Metcalf, 2007). A consensus view on the marginal damages of greenhouse gases does not exist (Metcalf, 2007).

2. R/ECON™ – The New Jersey State Economic Model

R/ECON™ is an econometric model comprised of more than 300 equations that are solved simultaneously. The equations are based on historical data for New Jersey and the United States. The historical data used to produce the model cover the period from 1970 to 2006, with some sectors updated through 2007. The sectors in the model include:

- Employment and gross state product for forty industries;
- Wage rates and price deflators for major industries;
- Consumer price index;
- Personal income and its components;
- Population, labor force, and unemployment;
- Housing permits, construction contracts, and housing prices and sales;
- Energy prices and usage;
- Motor vehicle registrations and stocks; and
- State tax revenues by type of tax, and current and capital expenditures.

The heart of the model is a set of equations modeling employment, wages, and prices by industry. In general, employment in an industry depends on demand for that industry's output, and on the state's wages and prices relative to the nation's wages and prices. Demand can be represented by a variety of variables, including (but not limited to) New Jersey personal income, population, sectoral output, or U.S. employment in the sector. Growth in population is driven by total employment in the state and by state prices relative to national prices.

As part of the New Jersey Energy Master Plan effort, the model was extended to include additional equations related to the energy sector. The equations in this new model sector are:

- Electric prices, usage, and revenue for residential, commercial, industrial, and other consumers;
- Natural gas prices, usage, and revenue, for residential, commercial, and industrial users and the electric power sector;
- Fuel oil prices and usage for residential, commercial, industrial, and other consumers
- Motor fuel and diesel prices and usage;
- Energy sales and business taxes in millions of dollars; and
- Employment at electric utilities and other utilities.[13]

The R/ECON™ forecasting service produces four forecasts of the New Jersey economy each year. The final modeling for the New Jersey Energy Master Plan used the June 2008 R/ECON™ forecast as its baseline for the BAU and the June 2008 R/ECON™ Pessimistic forecast for the BAU Pessimistic baseline. Both baseline forecasts go out to 2020. The data

13 The employment data, like all other New Jersey employment data used in the model, come from the New Jersey Department of Labor.

used for the United States came from Global Insight, Inc., a national leader in economic forecasting.

D. The Business as Usual and Energy Master Plan Scenarios

In developing the New Jersey Energy Master Plan, four major scenarios were evaluated during the Plan's time horizon, which ends in the year 2020. The BAU scenarios represent two specific possible economic futures: a "more likely" scenario and a pessimistic scenario in which energy price are higher than in the "more likely" scenario. These BAU scenarios assume that New Jersey continues on its current course with respect to energy policy.

The Energy Master Plan scenarios utilize a combination of programs designed to implement energy efficiency, renewable energy, and other actions intended to achieve the greenhouse gas emission limits specified in the New Jersey Global Warming Response Act and the RPS requirements for 2020.

1. Comparison of BAU and Energy Master Plan Electricity Assumptions

This section describes briefly the major similarities and differences between the BAU and Energy Master Plan scenarios. Most of these policies are assumed to be implemented on January 1, 2010; they escalate linearly until they achieve their final level on December 31, 2020.

Demand Growth – the BAU scenarios assume that electricity demand growth reflects the R/ECON™ June 2008 trajectory, averaging about 1.27 percent annual sales growth across all sectors. The PJM load forecast assumes load to grow at a specific percentage in each zone by year (PJM, 2007).[14] The Energy Master Plan scenarios, however, assume that much of this increase in demand is met through energy efficiency and demand response measures that are not part of the BAU scenario. The Energy Master Plan scenarios also assume that 900 MW can be shaved off the top fifty peak demand hours through the use of demand response.

Renewable Generation – both scenarios assume the same Renewable Portfolio Standard (RPS). Since the RPS is based upon a percentage of demand, the amount of renewables in the BAU scenario is greater than in the Energy Master Plan scenario. Compared to the BAU scenario, the Energy Master Plan scenario assumes that New Jersey installs an additional 2,650 additional MW of off-shore wind, 100 additional MW of on-shore wind, 450 MW of additional biomass, and 50 MW of new and emerging technologies.

Combined Heat and Power (CHP) – the Energy Master Plan scenario assumes an additional 1,500 MW of CHP is implemented behind-the-meter by 2020.

Conventional Generation – Differences in conventional generation are based upon the differences in demand and generation between the two scenarios.

New Transmission – New transmission is identical for all scenarios and includes the specific additions to the existing PJM transmission system.

14 See Jay Apt, Lee Gresham, M. Granger Morgan, and Adam Newcomer, Incentives for Near-Term Carbon Dioxide Geological Sequestration: A White Paper prepared for The Gasification Carbon management Work Group, Carnegie Mellon Electricity Industry Center, Oct. 9, 2007 who point out that national electricity load growth is linear not exponential.

2. Comparison of BAU and Energy Master Plan Scenarios Natural Gas and Fuel Oil Assumptions

Natural gas and fuel oil are considered in two parts: space heat and non-space heat. Under the State's greenhouse gas mitigation policies, savings through energy efficiency are expected for non-space heat natural gas and fuel oil consumption. Only the heating sector will be affected by the Plan.

Demand Growth – The BAU scenario assumes that demand for natural gas remains steady overall at 490 trillion Btus. There will likely be a slight decline in consumption in the residential sector, while the commercial sector will likely see a significant increase in gas use. Distillate Fuel oil, which has been declining in favor of natural gas for many years, decreases from approximately 80 trillion Btus in 2004 to 30 trillion Btus in 2020. In the Energy Master Plan scenario due to energy efficiency gains, demand for both natural gas and fuel oil is projected to decline to about 470 trillion Btus.[15]

Energy Efficiency – The assumed decline in consumption is driven by four policies: implementation of the EPAct 2005 appliance standards; adoption of future appliance standards; adoption of enhanced building codes for new construction; and development of an energy efficiency incentive program (in the form of white tags or rebates) targeted specifically at existing building stock.

E. Modeling Results

1. Electricity, Natural Gas, and Fuel Oil Results

a. Electricity

This section presents wholesale electricity results for the two main scenarios. Table 1 compares the wholesale electricity prices in the BAU and Energy Master Plan scenarios in the year 2020. Prices reported are the straight hourly averages of New Jersey electricity prices ($/MWh), the load weighted average New Jersey electricity price, and the installed capacity price. Similar calculations are performed for Class 1 renewable energy resources such as solar and on- and off-shore wind installed in New Jersey.[16] The delta in electricity prices between the Energy Master Plan and BAU scenarios are provided to the R/ECON™ model.

The LMPs, both straight average and load weighted, decrease between 7 and 8 percent between the BAU and Energy Master Plan scenarios. The capacity prices do not change in any of the scenarios because the marginal unit in all cases is a gas turbine that rarely, if ever, runs in the DAYZER simulation.[17] All renewable energy adders increase between the BAU

---

15 This reduction includes all savings from proposed energy efficiency policies proposed by the EMP, as well as gas needed to meet the proposed additional MWs of Combined Heat and Power (CHP).

16 While this report was being finalized, the NJ BPU issued rule changes to its solar programs. See Summit Blue Consulting, An Analysis of Potential Ratepayer Impact of Alternatives for Transitioning the New Jersey Solar Market from Rebates to Market-Based Incentives prepared for New Jersey Board of Public Utilities, December 6. 2007, and State of New Jersey Board of Public Utilities, Decision and Order Regarding Solar Electric Generation,Docket No. EO06100744, September 12, 2007.

17 Many stakeholders noted the rising costs of power plants and the belief that these increases were not just a short-term phenomenon. See Chupka and Basheda, Rising Utility Construction Costs: Sources and Impacts, The Brattle Group, Prepared for The Edison Foundation, September 2007.

and Alterative scenarios because the prices these resources obtain in the energy market decrease between these cases.[18]

Table 1. 2020 Business As Usual and Energy Master Plan Nominal Electricity Prices ($/MWh)

Prices	BAU-National	Energy Master Plan-National
Straight Avg. LMP ($/MWh)	114.59	107.03
Load Weighted Avg LMP ($/MWh)	119.31	110.42
Capacity Price ($/MWh)	18.79	18.79
Class I REC Price ($/MWh)	13.17	14.07
NJ On-Shore Price Adder ($/MWh)	19.34	20.23
NJ Off-Shore Price Adder ($/MWh)	55.04	58.56
SREC Price ($/MWh)	172.61	173.59
Emerging Techs. Adder ($/MWh)	N/A	36.69

Table 2 reports air emissions for 2015 and 2020 for the two major scenarios. In 2015, CO_2 emissions from New Jersey generators are greater in the Energy Master Plan than in the BAU scenario because of CHP emissions, although total CO_2 emissions in the Energy Master Plan scenario are less than total BAU CO_2 emissions because of a reduced reliance on imports. In 2020, CO_2 emissions are much greater in the BAU scenario than in the Energy Master Plan because the energy efficiency measures in the Energy Master Plan Scenarios are substantial enough to provide a counter-effect to the increased CHP emissions and there is a significant amount of electricity exported in the 2020 Energy Master Plan scenario.

Table 2. 2015 and 2020 Business As Usual and Energy Master Plan Scenarios (RGGI) Emissions

	2015 BAU	2015 ALT	2020 BAU	2020 ALT
SO_2 (Metric Tons)	38,437	45,008	64,385	59,439
NOx (Metric Tons)	18,632	20,209	30,489	27,311
NJ CO_2 (Million Metric Tons)	12.7	16.06	22.77	25.64
Imported CO_2	13.07	7.94	3.2	0
Total CO_2	25.77	24.01	25.97	25.64

18 The various REC prices and wind adders should not be added to the wholesale energy and capacity prices to arrive at the total wholesale price since these resources are only a fraction of the total MWhs sold.

B. Natural Gas and Fuel Oil Results

Because the natural gas and oil markets are continental and international, New Jersey is far less able to influence wholesale prices in the natural gas market with policies chosen under the Energy Master Plan than it is with electricity. Nevertheless, consumption levels and greenhouse gas emissions are affected in the Energy Master Plan scenario compared to BAU. Table 3 compares the greenhouse gas emissions from natural gas and fuel oil by sector in the BAU and Energy Master Plan scenarios. The reductions in emissions reflect reductions in consumption in all sectors, despite the significant increase in natural gas used for CHP under the Energy Master Plan Scenario.[19]

Table 3 shows the relative impact of the proposed Energy Master Plan policies on overall greenhouse gas emissions reductions. The biggest impact will come from the energy efficiency measures in existing building stock.

Table 3. 2020 Business As Usual and Energy Master Plan Scenario Greenhouse Gas Emissions from Natural Gas and Fuel Oil Combustion in the Residential, Commercial, and Industrial Sectors (million metric tons)20

Retail Prices	2004 Baseline	BAU-Natural Gas & Fuel Oil	ALT-Natural Gas & Fuel Oil	% Difference
Residential				
Space Heating	13.6	8.2	5.8	-29%
Other Combustion	3.9	3.5	3.3	-6%
Commercial				
Space Heating	6.6	8.0	5.6	-30%
Other Combustion	4.8	5.1	5.0	-2%
Industrial/Other				
Space Heating	0.9	0.6	0.6	-0%
Other Combustion	17.1	16.0	15.1	-6%

The New Jersey Energy Master Plan has significant impacts on all sectors of greenhouse gas emissions in New Jersey. Many New Jersey Energy Master Plan initiatives have direct and indirect effects on greenhouse gas emissions. There is an overall reduction of approximately thirty-three percent of greenhouse gases in the Energy Master Plan scenario compared to the BAU in 2020. The majority of the reduction is due to reduced demand in the electricity sector and reduced emissions from on-road gasoline as a result of New Jersey Energy Master Plan programs.

19 Reductions of 20 percent are based on energy efficiency policies, such as appliance standards, enhanced building codes, and programs targeting energy efficiency investments in existing building stock; the 9 percent reduction in the industrial sector recognizes the use of natural gas in production processes that may not have potential for efficiency gains.

20 Reductions in Greenhouse gas emissions were estimated by the NJ Department of Environmental Protection (DEP) and the Center for Climate Strategies, draft 10/15/07.

2. R/ECON™ Results

Assumptions pertaining to the BAU and Energy Master Plan Scenario cases were agreed upon by Energy Master Plan committee members, stakeholders, and the Governor's Office of Economic Growth. With these decisions made, the output from DAYZER and other sensitivity analyses are input into the R/ECON™ model to capture the macroeconomic impacts of the proposed policies. As noted above, the R/ECON™ model builds off a core model of more than 300 simultaneous equations to provide a multifaceted picture of the state's economy. The equations in the model are estimated using historical time-series at both the state and the national level for employment sectors, wages, consumer price index, population, tax revenues and expenditures, energy consumption, and retail energy prices. Projections of national trends are provided on a quarterly basis by Global Insight. Based on the estimated relationships between the state and national historical trends, projections are produced for the state's economy. However, assumptions about the state's economy that are different from the Business As Usual trends can be specified, and the effects of these changes—that is, how they ripple through the New Jersey economy—can be observed.

For the EMP, the additional cost to implement each proposed policy was translated into a unit that was then added to a specific variable in the R/ECON™ model. For example, the price impact of the proposed off-shore wind pilot project would be calculated first by determining the marginal difference in the cost of generating a unit of electricity compared to the average unit of electricity under the given generation fuel and technology mix. These calculations are determined by DAYZER. This per unit incremental cost is then multiplied by the total number of units expected to be produced in each given year. The total cost of this electricity generation is not paid by a single source, but rather spread over all electricity rate payers in the state. The final calculation provides a $/kWh value that is added to the R/ECON™ scenario to which the policy pertains; it is understood to be the incremental cost of implementing the off-shore wind pilot project. Similar calculations for approximately thirty proposed policies touching electricity, natural gas, and space heating complete the bridge between DAYZER, spreadsheet analyses, and the R/ECON™ model. Table 4 shows an example of these final adders to the R/ECON™ baseline data.

Table 4. Sample Adders for the Normal and Pessimistic Business as Usual Case

	Net Adder or (Subtractor)			R/ECON™ Variable affected
	2010	2015	2020	
Electricity (RPS and pilot project implementation)				
Normal BAU All Sectors $/kWh Adder	$ 0.00121	$ 0.00355	$ 0.00503	Retail Electricity price
Pessimistic BAU All Sectors $/kWh Adder	$ 0.00094	$ 0.00343	$ 0.00481	Retail Electricity price

Source: R/ECON™ adders supplied 09/25/2008. These adders fed into model output generated on 09/30/2008.

Table 5 shows some of the adders to the R/ECON™ model for the Energy Master Plan Scenario. In the cases where there are BAU adders, the value is *in addition to* the BAU.

The adders provide a bridge between DAYZER output, policy and cost assumptions not related to electricity, and the macroeconomic output provided by the R/ECON™ model.

Other policies such as the Combined Heat and Power (CHP) program proposed by the Energy Master Plan Committee have been handled in a different manner. CHP does not reduce electricity consumption beyond the explicit energy efficiency programs in the EMP, but it does reduce the electricity consumed from traditional power plants. To capture the 1500 MW of CHP installed and in use by 2020, the R/ECON™ model reduced the tax revenue generated from electric utility operation by the same percentage that the demand for traditional electricity is replaced by onsite, behind-the-meter-CHP generated electricity. This percentage starts at about 1.2 percent in 2010 (controlling for existing CHP assumed to be reflected in the data already) and increases to about 17 percent in 2020.

Aggressive energy efficiency and renewable energy policies, such as large in-state pilot projects for off-shore wind, on-shore wind, and biomass, imply new job creation in New Jersey. Jobs in the construction, operation, and maintenance of these pilot projects and employment requirements for improving energy efficiency in existing buildings in the state have been estimated and added to the model. The additional jobs from these "new" or expanded industries in the state largely offset the impacts on employment declines as a result of higher energy prices or lower demand for energy.

Table 5. Sample Adders for the Energy Master Plan Scenario

	Net Adder or (Subtractor)			R/ECON™ Variable affected
	2010	2015	2020	
Electricity (Implementation of RPS, more extensive pilot projects, and energy efficiency policies)				
Residential $/kWh Adder	$0.00	$0.00	$0.01	Retail Residential electricity price
Commercial $/kWh Adder	$0.00	$0.00	$0.00	Retail Commercial electricity price
Industrial $/kWh Adder	$0.00	($0.00)	$0.00	Retail Industrial electricity price
Natural Gas (Implementation of energy efficiency policies)				
Residential $/thousand cubic feet Adder	$0.17	$1.24	$2.31	Retail Residential Natural Gas price
Commercial $/thousand cubic feet Adder	($0.01)	$0.33	$0.39	Retail Commercial Natural Gas price
Industrial $/thousand cubic feet Adder	($0.01)	$0.30	$0.33	Retail Industrial Natural Gas price
Fuel Oil (implementation of energy efficiency and biofuel replacement policies)				
Residential $/gallon Adder	$0.00	$0.01	$0.01	Retail Residential Fuel Oil price
Commercial $/gallon Adder	$0.00	$0.01	$0.01	Retail Commercial Fuel Oil price
Industrial $/gallon Adder	$0.00	$0.01	$0.01	Retail Industrial Fuel Oil price

Source: R/ECON™ adders supplied 09/25/2008. These adders fed into model output generated on 09/30/2008.

Table 6 shows the range of retail prices given the adders and the predicted change in consumption. The difference between the "Baseline" and the "Energy Master Plan" is not equal to the adders simply because the model uses projection data for national wholesale prices from Global Insight, and the interactions between consumption and price changes may further have an impact on the price in 2020. However, the direction of the price increase is the same as anticipated from the adders.

Table 6. Retail Energy Prices Based on R/ECON™ Output

	2020 BAU	2020 Alt. Scenario	% Difference
Electricity			
Residential Price (Cents per KWH)	19.6	20.9	6.9%
Commercial Price (Cents per KWH)	17.0	17.7	3.9%
Industrial Price (Cents per KWH)	14.4	14.5	1.1%
Natural Gas			
Residential Price ($ per TCF)	25.31	27.65	9.3%
Commercial Price ($ per TCF)	21.65	22.04	1.8%
Industrial Price ($ per TCF)	18.51	18.84	1.8%
Electricity Price ($ per TCF)	13.87	13.87	0.0%
Heating Oil			
Residential Price (Cents per Gal)	304	305	0.3%
Commercial Price (Cents per Gal)	263	264	0.4%
Industrial Price (Cents per Gal)	241	242	0.4%

Source: R/ECON™ model output generated on 9/30/2008 (BAU) and 10/10/2008 (New Jersey Energy Master Plan).

The New Jersey Energy Master Plan focuses heavily on reducing energy consumption through enhanced energy efficiency measures. Table 7 shows the overall energy consumption reduction by implementing targeted programs. Total natural gas usage decreases only 15 percent between the BAU and Energy Master Plan scenarios because of the increase in natural gas consumption for behind-the-meter CHP units. Natural gas usage in the space heating and cooking sector decreases 27 percent between the BAU and Energy Master Plan, which achieves the goals of the New Jersey Energy Master Plan.

Table 7. Energy Consumption by Fuel Based on R/ECON™ Output

	2020 BAU	2020 Alt. Scenario	%Difference
Annual Use in Trillion BTUs: Total	2,028	1,782	-12%
Electricity (all sectors)	323	255	-21%
Total Natural Gas	636	540	-15%
Residential, Commercial, and Industrial Usage	478	351	-27%
Behind-the-Meter CHP Usage	12	105	759%
Natural Gas for Electricity	146	84	-42%
Fuel Oil (Distillate No. 2 and Residual)	303	219	-28%

Source: R/ECON™ model output generated on 9/30/2008 (BAU) and 10/10/2008 (New Jersey Energy Master Plan).

The impact of higher prices and lower consumption is that the overall sectoral and per customer expenditures decline between the BAU and the Energy Master Plan scenario.

Energy expenditures based on the R/ECON™ simulations indicate that the provisions of the EMP as quantified in the Energy Master Plan would lead to a savings of 30 percent for electricity and 22 percent for natural gas by 2020.

The following charts show the trajectory of energy usage in New Jersey in the Energy Master Plan scenario compared to the BAU. Even without the EMP, energy use per household and for the overall economy becomes more efficient. However, as the various provisions of the EMP go into effect, the difference between energy usage in the BAU and Energy Master Plan scenario increases, so that by 2020 energy usage both per household and as a proportion of Gross State Product is estimated to be 12 percent less in the Energy Master Plan than in the BAU.

Chart 1 shows the residential energy usage per household from 2007 through the modeling period of 2020.

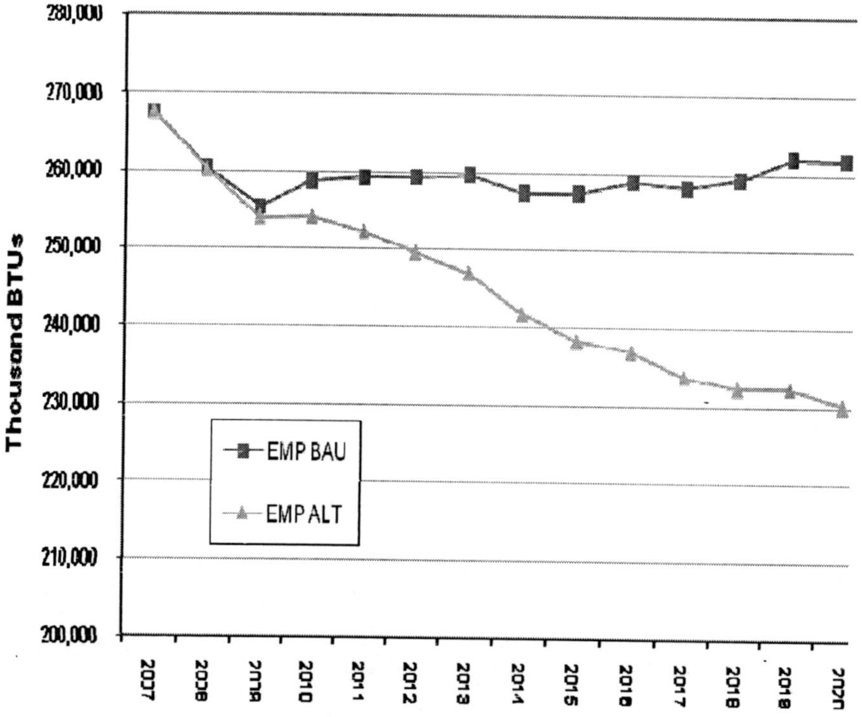

Chart 1. Residential Energy Use per Household, Baseline Compared to Energy Master Plan Scenario.

Source: R/ECON™ 09/30/2008 Output.

Chart 2 shows New Jersey's energy use per dollar of real gross state product from 2007 through the modeling period of 2020.

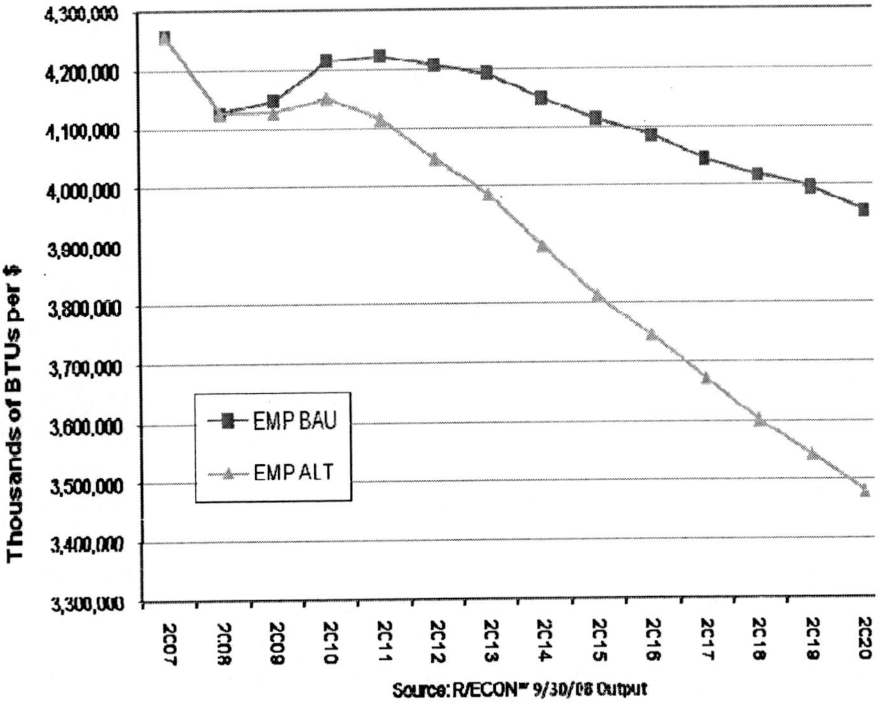

Chart 2. New Jersey Energy Use per Dollar of Real Gross State Product, Baseline Compared to Energy Master Plan Scenario.

Source: R/ECON™ 09/30/2008 Output.

Table 8. Macroeconomic Indicators Based on R/ECON™ Output

	2020 Average BAU	2020 Average Alt.	% Difference
Non-ag. Employment(thous)	4392.1	4410.7	0.4%
Unemployment Rate(%)	4.8%	4.7%	-0.8%
Personal Income($bill)	$791.0	$804.8	1.7%
Real Personal Income($bill, 2000)	$274.0	$278.5	1.6%
Retail Sales($bill)	$270.3	$274.0	1.4%
Real Retail Sales($bill, 2000)	$93.6	$94.8	1.3%
New Vehicle Registrations(thous)	658.8	659.0	0.0%
New Car Registrations	397.9	398.0	0.0%
New Light Trucks and Vans	260.9	261.0	0.1%
Residential Building Permits	26,204	25,466	-2.8%
Contract Construction($mill)	$14,818	$15,156	2.3%
Consumer Price Index(1982=100)	288.6	289.0	0.1%
Gross State Product($2000 bill)	$507.0	$507.4	0.1%
Total Tax Revenues($bill)	$51.2	$52.1	1.9%

Source: R/ECON™ model output generated on 9/30/2008 (BAU) and 10/10/2008 (New Jersey Energy Master Plan).

The macroeconomic effects of changes in energy prices and consumption are seen in Table 8. Most of the effects of the New Jersey Energy Master Plan policies are marginal. However, they do indicate that the EMP policies lead to an increase in non-agricultural employment (approximately 18,600 jobs), a decline in the unemployment rate, a 1.7 percent increase in personal income, and a 1.9 percent increase in major state tax revenues. As noted in Section III, the results below do not include the economic benefits of reducing greenhouse gases in the Energy Master Plan Scenario. Thus, even without accounting for the greenhouse gas reduction, the economy improves slightly under the Energy Master Plan Scenario as compared to the Baseline.

IV. Conclusion

We find that the process of creating an energy master plan requires a great amount of interaction between State leaders, stakeholders, and the general public. In New Jersey, such a comprehensive plan was possible for three reasons. First, strong executive leadership kept the New Jersey Energy Master Plan at the top of the list of policy priorities. Accompanying this leadership is the resources and commitment of the State. Second, a broad and inclusive process engaged key stakeholders and citizens and helped build support for the plan. The process in New Jersey was extensive and fluid, allowing for a number of diverse interactions between the State and stakeholders. Finally, the State engaged an academic team to develop a comprehensive modeling platform designed to analyze both the energy and macroeconomic impacts of the plan. Modeling provides a level of analytical legitimacy, which can help persuade critics that the plan will not harm the economy as the state pursues environmental goals. The New Jersey Energy Master Plan shows that investing in a cleaner energy future may in fact be beneficial to the economy, not harmful as some might claim.

Acknowledgments

The authors would like to acknowledge the financial support of the New Jersey Board of Public Utilities and the comments and suggestions of Lance Miller and Bharat Patel of that body. Erin Coughlin of the Center for Energy, Economic and Environmental Policy provided research support.

References

American Planning Association (2004). American Planning Association Policy Guide for Energy.

Andrews, C. J. (2008) Energy conversion goes local. *Journal of the American Planning Association,* vol. 7 (no. 2), pp. 231-254.

Andrews, C. J. (2006). Sustainable energy in New Jersey. In D. Rahm (Ed.), Sustainable Energy and the States: Essays on Politics, Markets and Leadership (pp. 108-126). Jefferson, N.C.: McFarland & Company, Inc.

Andrews, C. J., Jonas, H.C., Mantell, N., & Solomon, R. (2008). Deliberating on statewide energy targets. *Journal of Planning Education and Research.* vol. 28 , pp. 6-20.

New Jersey State Legislature (2007). ASSEMBLY [Bill], No. 3301 "Global Warming Response Act" available at http://www.njleg.state.nj.us/2006/Bills/ A3500/3301_ R2.HTM.

California Energy Commission (2003). 2003 Integrated Energy Policy Report, 100-03-019. Sacramento, CA: California Energy Commission.

Center for Energy, Economic and Environmental Policy (2008), Modeling Report for the New Jersey Energy Master Plan. Available at http://www.state.nj.us/ emp/docs/pdf/10122208ceeepModEMP.pdf.

Department of Environmental Protection and Energy (1991). 1991 New Jersey Energy Master Plan, November 1991. Trenton, NJ: State of New Jersey.

State of New Jersey (2007). Executive Order 54 available at http://nj.gov/infobank/circular/ eojsc54.htm.

Florida Department of Environmental Protection (2006). *Florida Energy Plan.* Tallahassee, FL: State of Florida.

Intergovernmental Panel on Climate Change (2007)., "Climate Change 2007: The Physical Science Basis," *Contribution of Working Group I to the Fourth Assessment Report of the IPCC*, Eds. S. Solomon, D. Qin, M. Manning, M. Marquis, K. Averyt, M. Tignor, H. Miller, Z. Chen Cambridge, UK: Cambridge University Press

Kann, A. & Wyant, J. (2000). Approaches for performing uncertainty analysis in large-scale energy/economic policy models. *Environmental Modeling and Assessment,* vol. 5, pp. 29-46.

Lee, R. (2002). The Environmental impacts of energy use. In R. Bent, et al. (Eds.), *Energy: science, policy, and the pursuit of sustainability.* Washington, D.C.: Island Press.

Metcalf, G. E. (2007). The Economics of climate change. Cambridge, UK: Cambridge University Press.

Metcalf, G. E. (2007). A Proposal for a U.S. Carbon Tax Swap: An Equitable Tax Reform to Address Global Climate Change, The Brookings Institution, Oct. 2007 citing the Intergovernmental Panel on Climate Change (IPCC), Contribution of Working Group II to the Fourth Assessment Report, Geneva, Switzerland, 2007.

New Jersey Board of Public Utilities, New Jersey Energy Master Plan Committee (1995). New Jersey Energy Master Plan: Phase I Report. Trenton, NJ: State of New Jersey.

New Jersey Board of Public Utilities (2006). Section 1: introduction, proposed goals and institutional process. Available at http://www.nj.gov/emp/home/docs/ approved/ 060929.html.

New Jersey State Legislature (1987). NJ State Law, 52:27F-14. Energy Master Plan Committee, L. 1977, c. 146; amended 1987, c. 365, s. 14.

New York State Research and Development Authority (2002). 2002 State energy plan and final environmental impact statement. Albany, NY: State of New York.

Park, H. & Andrews, C. J. (2004). City planning and energy use. In C. J. Cleveland (Ed.), Encyclopedia of Energy (Volume 1, pp. 317-330). Elsevier.

PJM Capacity Adequacy Planning Department (2007). PJM Peak Load Forecast. Norristown, PA: PJM.

U.S. Energy Information Administration (2009). How dependent are we on foreign oil: Energy in Brief. Available at http://tonto.eia.doe.gov/energy_in_brief/foreign_oil_ dependence.cfm.

Vermont Department of Public Service (1998). Fueling Vermont's future comprehensive energy plan and greenhouse gas action plan – overview (Pursuant to 30 V.S.A. §202b). Montpelier, VT.

West Virginia Division of Energy (2007). West Virginia energy opportunities: a blueprint for the future resources for economic growth and energy security. Available at http://www.energywv.org/community/EOD.pdf.

In: Environmentalism: Environmental Strategies...
Editors: E. Kahraman and A. Baig, pp.73-99
ISBN: 978-1-60741-634-0

Chapter 3

REMOVAL OF HAZARDOUS MATERIALS FROM AQUEOUS WASTE STREAMS USING BIOMASS AND ENZYMES WITH REFERENCE TO ENVIRONMENTAL CONTROL: REVIEW

Fatma Gurbuz*
Division of Biological Sciences, University of Suleyman Demirel,
32360 Isparta, Turkey

ABSTRACT

Removal of toxic and heavy metal contaminants from aqueous environments is one of the most important environmental issues to face the world. Conventional treatment technologies (chemical precipitation, ion-exchange resins, membrane separation, etc.) for removal of toxic substances from aqueous solution are not economical and also generate huge quantities of toxic chemical sludge. Hence, research has been intensified on the search for alternative and innovative wastewater treatment techniques. The techniques mostly have focused on microorganisms. Metabolically active biomasses and more complex systems such as employing enzymes are considered to be potential instruments for removal of these pollutants from waste effluents. Attention has been paid recently to the enzyme technology. A large number of enzymes from a variety of different microorganisms have been reported to play an important role in an array of waste treatment applications. However there are challenges to overcome in these systems due to sensitivity to extremes of pH, high concentration of pollutants, and requirement of external metabolic energy. Hence the isolation and selection of resistant strains are crucial in order to overcome the limitations of employing living cells. Studies on bioaccumulation have shown that some strains of bacteria, algae (microalgae and cyanobacteria) and fungi isolated from contaminated sites possess excellent to survive capability and have high tolerance to various toxic metals. These microorganisms may be promising candidates for detoxifying and removing toxicants from wastes.

* Fax: +90-246-2371106; E-mail: fatma_gurbuz@yahoo.com, fgurbuz@fef.sdu.edu.tr

In this review aerobic degradation through bioaccumulation by bacteria and microalgae and enzyme-catalyzed reduction-based remediation of toxicants from waste waters are discussed.

Keywords: Biological treatment, bioaccumulation, enzyme, environment, microorganism.

1. Introduction

Industrial, agricultural, and residential effluents enter our waterways polluting these systems with a wide range of organic, metallic and inorganic compounds. Hence the delicate balance of our planet's fragile aquatic ecosystem is being disturbed at an alarming rate. Intensification of industrial activities in recent years has resulted in contributes to an increasing dispersion of toxic compounds and chemical substances into natural environments mainly to aquatic systems. Especially waste streams from mining; plastics; electroplating; leather tanning; galvanizing, pigment and dyes manufacture; metallurgical and paint industries; and other metal processing and refining operations contain metal ions and toxic, hazardous chemicals. These wastes brought devastating and permanent contamination to the environment due to their non-degradable and persistent nature (Volesky vd. 1993, Wilde and Beneman 1999). These toxic metal ions are not only potentially hazardous to human health but also to other life forms. Toxic metal ions may sometimes cause life-threatening illnesses due to irreversible effects on vital tissues (Malik, 2004).

The pursuit for a clean environment creates the need to develop industrial wastewater treatment methods with better performance efficiencies than conventional methods of physico-chemical processes such as oxidation and reduction, chemical precipitation, filtration, electrochemical treatment, evaporation, ion-exchange and reverse-osmosis processes. The choice of either biological or chemical methods for treatment of industrial wastewaters depends on the composition of the wastes. Bioprocesses are systems in which complete living cells (bioaccumulation) and their enzymes are used to effect desired physical or chemical changes (Gadd.2000). Microorganisms have several uses in the environment. One application is in the control of pollution and treatment of toxic wastes and heavy metals. Additionally, techniques are beginning to be used to select micro-organisms that can degrade extremely toxic compounds (phenols and aromatic, amines, cyanide wastes). In the mining industry, microbes are used to leach metals from mine dumps and concentrate metals from dilute solutions, and there are possibilities for using biotechnology to improve the efficiencies of these processes. Metals are concentrated through a combination of surface reactions, intra and extracellular precipitation, and intra and extracellular complexation reactions (Veglio and Beoclhini 1997). Using growing cultures in biological removal could avoid the need for a separate biomass production process such as cultivation, harvesting, drying, etc., however there is need to maintain cell growth. The effectiveness of these processes is usually dependent on the parameters like temperature, toxicity, oxygen level and availability of nutrients etc. If the problem of metal toxicity to the growing cell is overcome by the use of metal-resistant organisms, the continually self-replenishing system can be left to run continuously for extended periods (Dursun 2003). The discussion here, therefore, focuses on the bioaccumulation and enzyme catalyzed processes in the treatment of liquid wastes which contain of heavy metals, toxic chemicals (cyanides and metallocyanides), and phenols.

2. Practical Uses for Enzymes

2.1. Use of Enzymes in Wastewater Treatments

Biological processes that convert one material into other using biological agents (e.g., living microbes or enzymes) involve biological remediation techniques that generally require either aerobic or anaerobic digestion. The performance of these standard aerobic and anaerobic biological treatment processes could conceivably be improved by the addition of specific enzymes that could augment the ability of the natural micro-organisms to degrade, for example, protein, starch, polysaccharides, and celluloses. With the necessity for improvement in biological remediation techniques, enzyme technology has been receiving increased attention. Enzymes are biological catalysts responsible for supporting almost every type of chemical reaction. They are physiologically important because they speed up, at least 1000-fold, the rate of reactions by decreasing the amount of energy required to form a complex of reactants (Whiteley and Lee 2006). The particular benefits offered by enzymes are specificity, mild conditions and reduced waste. It may be possible, by choosing the right enzyme, to control which products are produced, and unwanted side reactions are minimized due to specificity of enzymes that appear in the waste stream (Posorske 1984). They are less likely to be inhibited by substances which may be toxic to living organisms (Aitken 1993). Moreover enzymes operate over a broad aromatic concentration range and require low retention times with respect to other treatment methods (Siddique et al. 1993). Enzyme-based processes tend to have lower waste treatment costs. Such enzymes could be applied selectively at specific wastewater treatment plants where their particular substrates are present in unusually high concentrations.

The kinetics of simple enzyme catalyzed reactions was first characterized in 1912 by Michaelis and Menten who derived the hyperbolic equation 1:

$$v1 = \frac{V\max[S]}{Km + [S]} \tag{1}$$

where $v1$ is the rate of the enzyme catalysed reaction, $[S]$ is the concentration of substrate, Km the Michaelis–Menten constant and $V\max$ is the maximum reaction rate. It can be established that Km is also equal to the substrate concentration that would give 50% $V\max$. At low substrate concentrations, the rate of an enzymatic reaction is of first-order and is directly proportional to substrate concentration ($v = k[S]$) (Whiteley and Lee 2006).

Most enzymes normally exhibit a strong dependence of activity on the pH of the medium (Clark and Switzer, 1977); therefore it is important to optimize the pH of a particular enzyme. The optimal pH for hydrogenase activity for the treatment of textile effluents has been found to be in the range of 7.0–8.0 (Ngwenya, 2004). Optimum pH for an enzyme lies somewhere between the pKa values of the two (or more) catalytic amino acids. Any pH lower than the optimum would protonate the ionized residue while any pH higher than the optimum would ionize the protonated residue leading to decreased activity. pH dependence of hydrogenase reactions is the consequence of varying ionization of functional groups in the enzyme as well as the substrate (Rashamuse, 2003). Most enzymes have an optimum temperature to function. This may be related to the type of organism from that the enzyme was isolated. Some organisms grow well near room temperature and so their enzymes are most active at a temperature around 30–40 °C. At low temperatures, all the molecules in the solution slowdown and so does the enzyme catalyzed reaction while at high temperatures, the thermal movement of the molecules become too great for the enzyme to hold its shape or conformation and so it begins to denature and lose its activity.

Microbial enzymes are often more useful than enzymes derived from plants or animals. This is because; of availability of great variety of catalytic activities, possibility the high yields of products, ease of genetic manipulation and rapid growth of microorganisms on inexpensive media. Microbial enzymes are also more stable than their corresponding plant and animal enzymes and their production is more convenient and safer (Wiseman 1995). Recently, microorganisms and enzymes have been successfully used in diverse bioremediation applications (Gavrilescu, 2004; Jördening and Winter, 2004). Bacteria utilize a combination of extracellular and intracellular adaptive and constitutive enzymes in the biochemical breakdown and assimilation of the organic and inorganic compounds present in liquid wastes. These compounds, along with various nutrients (i.e. nitrogen, phosphorous and trace metals), are degraded and/or assimilated for the purposes of energy production and cell synthesis. Extracellular and intracellular enzymes are either adaptive or constitutive in the nature. Constitutive enzymes have evolved following many successive generations and mutations resulting from exposure of a bacterial species to specific compounds. The production of adaptive enzymes that are not normally present may be stimulated by the presence of a particular compound or class of compounds that the bacteria utilize as alternative carbon and/or energy sources if preferred substrates are not available (Gurbuz et al., 2004). For instance, *Bacillus* species have been shown to produce a number of enzymes including, cellulase (Mawadza et al., 2000; Singh et al., 2005), serine protease (Balaban et al., 2004), alkaline protease (Prakasham et al., 2005), xylanase (Moriya et al 2005), lipase (Hasan 2005) that may be utilized industrially.

Enzymes are increasingly penetrating the chemical industry as catalysts for numerous reactions. Various technologies which catalyzed by enzyme may be categorized into several divisions (figure 1). A large number of enzymes (peroxidases, oxidoreductases, cellulolytic enzymes, cyanidase, proteases, amylases, etc.) from a variety of different sources have been reported to play an important role in various waste treatment applications. In recognition of these potential advantages, recent research has focused on the development of enzyme-based processes for the treatment of wastewaters, solid wastes, hazardous wastes and soils (Karam and Nicell 1997). Enzymes are used in bioleaching of pulp to reduce chlorine consumption; pulp de-watering and de-inking; removal of pitch; degradation of dissolved and suspended organics in concentrated effluents of mills; and enhanced fibrillation to give stronger paper (Ericksson, 1997). Sangave and Pandit (2006) demonstrated that pre-treatment of distillery effluent with the enzyme cellulose increased the rate of COD removal. The feasibility of applying enzymes for the degradation of phenolic compounds has received much attention. Klibanov et al. (1980) employed horseradish peroxidase to remove phenols from wastewater. Horseradish peroxidase catalyzed hydrogen peroxide to oxidize phenols and generated phenoxy radicals, and then these radicals reacted to form oligomers and polymers, which were precipitable and could be readily removed from wastewater (Erhan et al., 2002). Lipases have been used for the degradation of wastewater contaminants from olive oil processing (De Felice et al., 2004). Lanciotti et al. (2005) evaluated the ability of several *Y. lipolytica* strains to grow in olive mill wastewater and to metabolize its lipid fraction. All strains studied were able to produce extracellular lipases. Some strains, in addition to a high lipase activity, induced a COD reduction with respect to the uninoculated wastewater ranging from 20% to 40% after 72 h at 25°C. Table 1 summarises different enzymes, which have been studied for their toxicant degradation.

Table 1. Summary of enzymes and their potential applications in waste treatments

Enzyme	Application	References
Cyanidase, cyanide dihydratase	Cyanide decomposition	(Basheer et al.,1993; Jandhyala et al 2003)
Laccases	Phenols	(Majcherczyk et al 1998), (d'Acunzo et al 2002)
	Aromatic compounds	(Nicell 1994 Sharma et al 2003)
Horseradish peroxidase	Phenol and *p*-chlorophenol	(Cooper and Nicell 1996; Lai and Lin 2005; Bayramoglu and Arıca 2008)
Lignin peroxidase	Polycyclic aromatic hydrocarbons	(Bogan and Lamar et al.,1996; Wu and Ozaki 1996)
Phosphatase	Removal of heavy metals, sulfate reducing	Macaskie et al 1987;Whiteley et al.,2002)
Parathion hydrolase	Hydrolyzation of organophosphate pesticides	(Coppella et al., 1990)
Cellulase Cellobio-hydrolase	Hydrolysis of cellulosic sludges from pulp and paper to produce sugars and alcohol	Duff et al 1995; van Wyk and Mohulatsi 2003.
Chromate reductase	Chromate degradation	(Park et al 2000)
Azo reductases	Azo dyes	(Russ et al., 2000)
Lipases;	Biodegradation of polymers	(Marten et al., 2003; Sivalingam et al., 2003)
Endo-xylanase	Degradation Kraft -pulp	(Bocchini et al .,2003, Kansoh et al., 2004)

Michael et al. (1991) and Jaspers et al. (1994) demonstrated that purified lignin peroxidase and manganese peroxidase were effective in the decolorization of kraft bleach plant effluent. Sayadi and Ellouz (1995) also observed the decolorization of olive mill waste waters by these two purified enzymes. Their study demonstrated that crude enzyme extract from *Serratia marcescens* AB 90027 was used to catalyze hydrogen peroxide to oxidize and degrade hydroquinone. In addition, the degradation of several phenolic compounds by such an enzyme/H_2O_2 treatment was evaluated No peroxidase activity was found in the extract. However it was suggested that the enzyme from *Serratia marcescens* AB 90027 might be a type of catalase which strengthened and accelerated this oxidation process (Yao et al 2006).

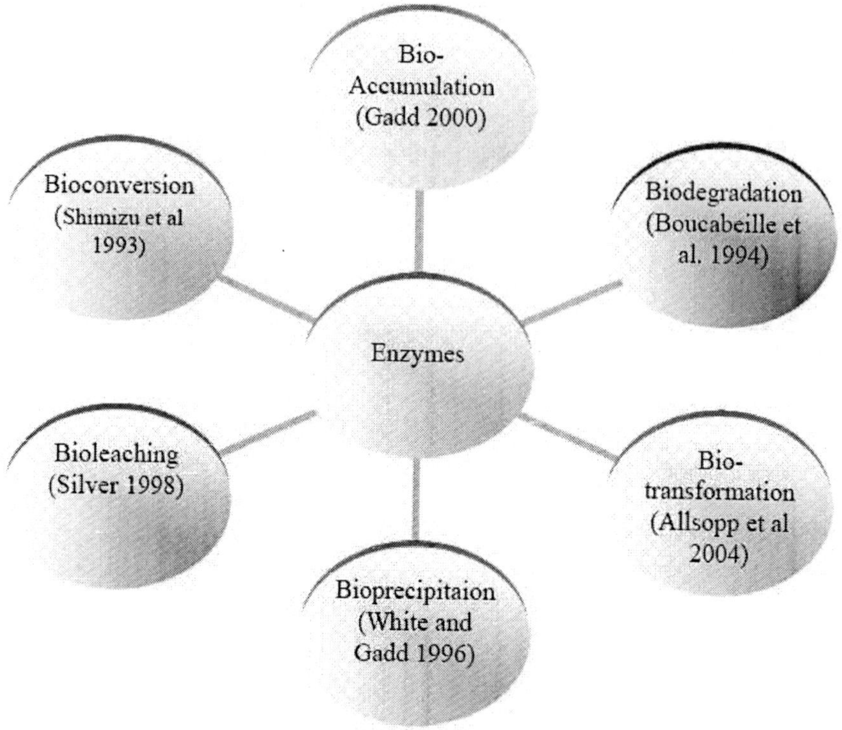

Figure 1. Various biological processes which involve enzymes.

2.2. Immobilisation of Biocatalyst and Uses of Enzymes as Biosensors

Despite their advantages, the use of enzymes in industrial applications has been limited by several factors, mainly the high cost of enzymes, their instability, and availability in only small amounts (Freire et al., 2002). Also, because the enzymes are soluble in aqueous media, it is difficult and expensive to recover them from reactor effluents at the end of the catalytic process. This restricts the use of soluble enzymes in batch operations, followed by disposal of the spent enzyme-containing solvent. To overcome this problem, biocatalysts have been immobilised (Farré et al., 2005). This technique offers the wider and more economical exploitation of biocatalysts in the waste treatment, medicine, and in the development of bioprocess monitoring devices. Immobolised enzymes in the food industry have a long history of use (Sharma and Messing, 1980). Biocatalysts can be immobilised using either isolated enzymes or the whole cells or cellular organelles. For instance, use of the linked enzyme system of urease and glutamic dehydrogenase has facilitated the measurement of Hg (II), Cu (II), Cd (II), and Zn (II) at concentrations in the ppm–ppb range. The chromate reductase activity of cytochrome C_3 from sulfate reducing bacteria was used to construct a Cr (VI) selective biosensor (Micheal et al., 2003). Polyphenol oxidase (PPO) was immobilised on a glassy carbon electrode to measure cyanide with a detection limit of 0.1 nM (Shan et al., 2004). There are several advantages to enzyme biosensors. These include a stable source of material, the ability to modify the catalytic properties or substrate specificity by means of genetic engineering, and catalytic amplification of the biosensor response by modulation of the enzyme activity with respect to the target analyte (Rogers 2006). There are also some limitations for enzyme-based biosensors with respect to environmental applications. A number of enzymes have been shown to be inhibited by toxic metals that contaminate the environment (Freire et al 2002)

Immobilization of whole cells has been shown to be a better alternative to immobilization of isolated enzymes (Tampion and Tampion, 1987). Doing so avoids the lengthy and expensive operations of enzyme purification, preserves the enzyme in its natural environment thus protecting it from inactivation either during immobilization or its subsequent use in a continuous system. Applications for immobilized microalgae in environmental aquatic research have been increased recently: novel immobilization techniques as well as the use of living microalgae as biosensors in electronic devices designed to measure toxicity of substances and effluents. These applications were demonstrated to be a very promising topic in biotechnology research (Moreno-Garrido, 2008). Biosensors of different types involving microalgal cells have been designed in order to detect pollutants in the environment. Lukavsky and Maršálek (1997) immobilized *S. capricornutum* in 2% agar in order to use it as a biosensor for Cr^6+ toxicity. *C. vulgaris* has also been used in optic biosensors to determine the toxicity of herbicides (Naessens et al., 2000) such as atrazine, simazine and diuron, commonly used in cereal cultures. Campanella et al (2000) studied the cyanobacterium *Spirulina subsalsa* which was immobilized on to an oxygen probe and placed in a flowcell configuration. This algal biosensor was reversibly responsive to heavy metals, triazine herbicides and carbamate insecticides. Mallick (2002), demonstrated the use of immobilized algae for nitrogen, phosphorus and metal removal from wastewater. Environmental uses of immobilized algal cells are not restricted to pollutant removal. These techniques have been recently used in field of toxicity measurement experiments (Moreira dos Santos et al., 2004; Moreira et al., 2006). An important part of the current work in the field of the toxicity measurement is, however, focused on the design of microalgal-based biosensors (Chouteau et al., 2004; Shitanda et al., 2005). Algal bacterial consortium has been discussed by Munŏz and Guieysse (2006). Authors reviewed the interactions between algae and bacteria in processes designed for the treatment of hazardous contaminants and noted that oxygen production by algae had improved degradation of substances that must be degraded aerobically.

Genetically engineered cell biosensors have been used for environmental monitoring. For example E. *coli* was transformed for environmental monitoring of benzene derivatives (Berno et al., 2004), arsenite and antimonite (Rothert et al., 2005), Hg (II) (Biran et al., 2003). An automated continuous toxicity test system was developed using a recombinant bioluminescent freshwater bacterium *(Janthinobacterium lividum* YH9-RC) that was modified with *lux*AB and optimized for toxicity tests using different kinds of organic carbon compounds and heavy metals. In a study, luminously modified cells exposed to continuous phenol or wastewater stream showed a rapid decrease in bioluminescence, which fell below detectable range within 1 min. (Cho et al 2004). Microbial toxicity test systems employing whole cell bioluminescence are generally categorized into two systems based on physiological response of luminescent bacteria to toxic chemicals, luminescence emission or reduction. Ungenetically modified bacteria also have been used as indicators of toxic compounds. One example is *Pseudomonas pudita* which has been used to measure low concentrations lead exposure (Ogunseitan et al 2000). Again *P. pudita* has also been used as a biosensor for benzene (Lanyon et al 2005).

3. Use of Enzymes for Bioaccumulation

3.1. Bioaccumulation Mechanisms

3.1.1. Background

Bioaccumulation is the process responsible for the uptake of metal ions by living cells which involves enzymatic degradation, biosorptive mechanisms, together with intracellular accumulation and bioprecipitation (Gadd 2000).

Biosorptive mechanisms occur through several types of chemical reactions with the constituents of the cells, would attract and sequester the metals in biomass. Functional groups, such as: sulphhydryl and carboxyl groups in proteins, acetamido groups of chitin, structural polysaccharides of fungi, amino, hydroxyls in polysaccharide and mainly carboxyls and sulphates in polysaccharides of marine algae that belong to the divisions Phaeophyta, Rhodophyta and Chlorophyta found to be involved in metal binding present in the structural polysaccharides, lipid and proteins of the different biological organism. The weakly acidic corboxyl groups (R- $COOH^-$) play significant role in ion-exchange in *Chlorella vulgaris* (Cho et al., 1994). The strongly acidic suphate group (R- OSO^3) of fucodian and carragen in sea weeds, *Sargassum* and *Cladophara* and amino group (R_2NH) of chitin and chitosan (R-NH_2) in fungi has also been examined (Tsezos and Volesky 1988). Peptidoglycan carboxyl groups are the main binding site for cations in Gram-positive bacterial cell wall with phosphate groups contributing significantly in Gram-negative species (Beveridge and Doyle, 1989; McLean et al., 2002). Golab and Breitenbach (1995) indicated that the carboxyl groups of the cell wall peptidoglycan of *Streptomyces pilosus* were responsible for the binding of copper. Also, amine groups are very effective at removing metal ions, as it not only chelates cationic metal ions, but also adsorbs anionic metal species or dyes via electrostatic interaction or hydrogen bonding. Kang et al. (2007) observed that amine groups protonated at pH 3 and attracted negatively charged chromate ions via electrostatic interaction.

Some biosorbents can bind and collect a wide range of heavy metals with no specific priority, whereas others are specific for certain types of metals (Volesky and Kuyucak, 1988) Cost effectiveness is the main attraction of metal accumulation, therefore biomass can come from

- industrial wastes which should be obtained freely
- organisms easily available in large amounts in nature;
- organisms with rapid growth especially cultivated or propagated for biosorption purposes.

for metal removal experiments, these to be taken into account (Regine, et al., 2000).

3.1.2. Temperature Effect on Bioaccumulation

Many recent studies reported different viewpoints for the effect of temperature on metal bioaccumulation. Some researchers had drawn the conclusion that metal ions accumulation was exothermic process, in which the metal removal rate decreased gradually as temperature increase (Li and Yuan, 2006). The others thought that the process was endothermic by their research (Salinas et al., 2000). Moreover, it was confirmed that Pb^{2+} sorption by *R. glutinis* was not affected by temperature (Cho and Kim, 2003). The increase of Cr (VI) accumulation with increasing temperature may be also due to the higher affinity of sites for Cr (VI) or an increase in the number of binding sites on relevant biosorbents surfaces as a result of reorientation of cell wall components. Similar results had been informed by Sultan and Hasnain (2007). Authours have found that Cr (VI) reduction was drastically reduced below 28°C and almost ceased at 20 °C. Reduction of toxic hexavalent chromium by *Ochrobactrum intermedium* strain SDCr-5 was found to be optimum at 37°C. Another study about Cr (VI) degradation by *Bacillus* sp. showed that optimum temperature for Cr (VI) reduction occurred at 30°C (Wang and Xiao, 1995). Arica et al. (2005) suggests that when the temperature increases promoted the removal rate increase because of the rise in kinetic energy of biocatalyst and also due to collision frequency increase between sorbent and sorbate.

3.1.3. pH Effect on Bioaccumulation

In general, increasing the pH increases the overall negative charge on the surface of cells until all the relevant functional groups are deprotonated, which favors the solution chemistry not only the bacterial surface chemistry, but the metal/dye speciation as well electrochemical attraction and adsorption of cations. Anions would be expected to interact more strongly with cells with increasing concentration of positive charges, due to the protonation of functional groups at lower pH values (Vijayaraghavan and Yun 2007). Metal ions in solution undergo hydrolysis as the pH increases. In the case of nickel removal from aqueous solution, López et al. (2001) indicated that within the pH range from 1 to 7, nickel existed in solution as Ni^{2+} ions (90%); whereas at pH 9, Ni^{2+} (68%), Ni_4OH_4 $^{4+}$ (10%) and Ni (OH)+ (8.6%) co-existed. Park et al (2007) suggested that when Cr (VI) comes in contact with biomaterials, especially at low pH, the Cr (VI) can be easily or spontaneously reduced to the Cr (III), because Cr (VI) has high redox potential value (above + 1.3 V at standard condition). The different chemical species of a metal may occurre with different pH. This will provide variable charges and absorbability at solid–liquid interfaces.

3.2. Bioaccumulation of Heavy Metals

Discharge of heavy metals from metal processing industries is known to have adverse effects on the environment. Hence heavy metals in water have long been of concern. What kinds of metals should we be interested in examining—in terms of removing their threat from the environment? The choice of metals to study for bioprocess mostly depends on interest and the impact of different metals. Metals on the basis of environmental threats, that is, toxic heavy metals and radio nuclides are mainly of interest therefore they been studied a great deal. Volesky (2007) suggests that metals known the metals are known for their high toxicity and impact: lead, mercury and cadmium should be of most concern because they have been connected to major poisoning or health hazards. These metals are closely followed by others, not quite as common, such as arsenic, (as arsenite) that has been a poison for centuries, and chromium, (as chromate) used in large quantities industrially for quite some time. As to radio nuclides, uranium and its uptake is popular subject to study because of environmental issues associated with it (Tsezos and Volesky, 1982, Marqués et al 1991).

Cadmium, Mercury and Lead Accumulation

Metal ions in the environment are bioaccumulated and biomagnified along the food chain. Mercury, lead, cadmium and chromium (VI) are defined as the most dangerous metals. Cadmium for instance has a half-life of 10–30 years and its accumulation in human body affects kidney, bone and also causes cancer (Moore and Ramamoorthy, 1984). It is a nonessential metal to living organisms and can become toxic by displacing zinc. Prasad et al. (1998) have shown that 11 mg/l cadmium linearly reduces chlorophyll a in *Clamydomonas reinhardtii* and at 22 mg/l cadmium; chlorophyll a was reduced in some cases by 76%. Similar results were obtained by Terry and Stone (2002). They have shown that *Scenedesmus abundans* was sensitive to cadmium although the alga removed the cadmium significantly. Chlorophyll *a* decrease occurred by the increasing concentration of Cd. Cadmium removal of capacity of living cells of the gram-negative bacterium *Sphingomonas paucimobilis* was found distinctly higher than that of nonliving cells. Nevertheless at this study also shown that bacterial growth was inhibited (Tangaromsuk et al., 2002). Immobilized live cell and heat inactivated of *Oscillatoria* sp. was studied in order to remove cadmium (II) ion from aqueous solution by Katırcıoğlu et al (2008). Authors have shown the immobilized live *Oscillatoria* sp. H1 revealed a higher biosorption capacity (32.2 mg g^{-1}) than heat-inactivated *Oscillatoria* sp. H1 (27.5 mg g^{-1}). However, there is also evidence to the contrary. Living and non living *P. aeruginosa* ASU 6a were exposed Pb^{2+} at different concentrations. Removal of lead ions from aqueous solutions were success however nonliving organism was performed better (Gabr et al., 2008).

Hg(II)-containing compounds are environmentally toxic due to both organomercurial compounds and inorganic forms of mercury, cross biological membranes and bind to thiol and amino groups in essential enzymes. The consequences of these biological interactions will cause of inactivation of enzymes (Kerper et al., 1992). The biosorption of Hg^{2+} by two strains of cyanobacteria, *Spirulina platensis* and *Aphanothece flocculosa*, was studied under a batch stirred reaction system. The maximum attainable uptake values of Hg^{2+} by *S. platensis* and *A. flocculosa* were 456.0 mgg^{-1} and 428.4 mgg^{-1} respectively (Cain et al., 2008). Alive bacteria are sensitive to mercury, whereas genetically engineered bacteria with both high bioaccumulation capacity and high affinity to desired metals such as mercury, lead and nickel. Experimental studies have showed the recombinant bacterial cells could selectively accumulate the goal metal ions from multi-component solutions under varying environmental conditions. For example, genetically engineered *E. Coli.* was demonstrated to accumulate mercury (Deng and Wilson 2001) and Cd^{2+} (Deng et al 2007) from aqueous solutions.

Uranium Accumulation

Uranium is a common element. Dissolved U is found in most natural waters at very low concentrations, but is of some concern in waters adjacent to U mining operations when concentrations increase to levels above 1 mg dm^{-3}. ^{238}U, the most common radionuclide, has a half-life of $4.5x10^9$ years, which makes clean-up environmentally important (Meinrath et al., 1996). Therefore U removal from waste streams has been tackled by intracellular sequestration by growing plant, algal and microbial cells. Among the microorganisms, algae are of predominant interest for the ecological engineer because of their ability to sequester U and because some algae can live under many extreme environments, often in abundance. *Scenedesmus* is a microalgae genus commonly used in heavy metal removal experiments. It has proven removal capacity for U^{6+} (Zhang et al., 1997). Recently removal of uranium from mining waste water using algal/microbial biomass has been shown by Kalin et al (2005).

Arsenic Accumulation

Natural contamination of groundwater with arsenic is a global health problem (Ayotte et al., 2003). Long-term oral exposure to arsenic in drinking water is reported to cause skin lesions, and skin, lung, bladder and kidney cancer (WHO, 2001). Arsenic in groundwater is mainly found in two oxidation states, +5 (As (V)) and +3 (As (III)). The main arsenic forms in groundwater are species of arsenic (V) (H_3AsO_4) and arsenous (III) acid (H_3AsO_3) (Cullen and Reimer, 1989, Gailer 2007). Freshwater and marine micro and macro algae have been found to take up and bioaccumulate arsenate as a phosphorus analogue during normal metabolism (Maeda, 1994). *Chlorella vulgaris*, a freshwater alga, has been shown to accumulate arsenic and transform the inorganic arsenic species into arsenosugars at different arsenate concentrations. (Murray et al 2003).Arsenic (III) removal is reported to occur by different mechanism than As (V) removal. Immobilized biomass of blue green algae, *Scytonema*, was observed to effectively remove As (III) from water (Prasad et al., 2006).

Arsenic is toxic to many bacteria. However some bacteria are resistant to arsenic either due to the presence of a strictly phosphate-specific transport system, which prevents the uptake of arsenate which is analogous to phosphate (Willsky and Malamy, 1980), or due to an efflux system mediated by the plasmid- or chromosomally-encoded *ars* operon (Cervantes et al., 1994; Cai et al., 1998). Kostal et al. (2004) engineered *E. coli* to over express ArsR, the regulatory protein of the *ars* operon that has a specific binding site available for arsenite. Takeuchi et al (2007) demonstrated that *M. communis* is the first non-genetically engineered potent arsenic accumulating bacterium. When incubated with 5mgAs l^{-1} of arsenate, *M. communis* accumulated 2290μg As g dw^{-1} of arsenic. Overall they have found that adsorption of arsenite (As(III)) was smaller than adsorption of arsenate, and the maximum adsorption value of 843 μgAs g^{-1} dw was obtained for arsenate (As(V)). Hattunen et al (2007) reported that aminated *Lactobacillus casei* DSM20011 was observed to remove As (V) but not As (III) in water. At the lowest As (V) concentration (100 μg/l) it removed only 38.1±9.0% at pH 7 and authors suggested that the process was not practical as only As (V) was removed

Chromium Accumulation

Chromium (VI) salts originate from industrial sources such as leather tanning, photographic-film making, wood preservation, car manufacturing, petroleum refining and agricultural activity (Nkhalambayausi-Chirwa and Wang, 2001). Removal of chromium from wastewaters is obligatory in order to avoid water pollution. Legislation by different governments demands the concentration of Cr in discharges be less than 0.5 mg l^{-1} irrespective of its oxidation state (Bishnoi et al., 2007). Bacteria can reduce chromate to the insoluble and less toxic Cr (III), and thus chromate bioremediation is of interest. A number of chromium resistant micro-organisms have been reported to detoxify hexavalent chromium, which include *Pseudomonas* sp., *Microbacterium* sp., *Desulfovibrio* sp., *Enterobacter* sp., *Escherichia coli*, *Shewanella* alga, *Bacillus* sp., etc., where the method of detoxification is periplasmic biosorption, intracellular bioaccumulation and biotransformation through direct enzymatic reaction or indirectly with metabolites (Camargo et al., 2003). Cr reduction is an enzymatically catalyzed reaction attributed to soluble proteins for some species. In a study by Ackerley et al (2003), the chromate reductase activities of two electrophoretically pure soluble bacterial flavoproteins -ChrR (from *Pseudomonas putida*) and YieF (from *Escherichia coli*)- were examined. They found that both reduced chromate efficiently to Cr (III).

Partially purified chromate reductase from *Pseudomonas ambigua* G-1 and *Pseudomonas putida* MK1 have indicated Cr (VI) reduction using NADH or NAD(P)H as a source of electrons. In *Bacillus* sp. ES29 chromate reducing activity was localized in the cell-free extract (CFE) which utilizes NADH as the sole electron donor (Pal et al., 2005). Similarly Thacker et al (2007) have shown that hexavalent chromium decreasing activity was required NADH as cofactor. In their study, the bacterial isolate (*Brucella* sp.) reduced Cr (VI) in a concentration range from 50 to 300 mg l^{-1}. Microbial Cr reduction is a complex process with multiple possible pathways and unstable redox intermediates with intermediate valence between the valences of the stable end members. Evidence was reported by (Myers et al., 2000; Daulton et al 2007).

Growing microalgae also has been shown capable of reducing toxic hexavalent chromium. The short-term accumulation of chromate by *Anabaena* sp. and *Synechococcus* sp. PCC6301 was rapid with relatively low level of sorption to the cell wall. The process was regulated by chromate concentration described by Freundlich adsorption isotherm. On decreasing pH, chromate accumulation increased in both the species. Over a longer growth period *Anabaena variabilis* was capable of reducing chromate (VI) to chromium (III) and accumulating it (Garnham and Green, 1995). Another study was conducted by Yewalkar et al (2003). In their study, Cr (VI) reduction by unicellular non-motile algae belonging to *Chlorella* spp.was shown to grow in the presence of 30 μM Cr (VI).

4. Bioremoval of Toxic Chemicals

4.1. Uses in Detoxification of Cyanides and Metallo-Cyanides

Cyanide is highly toxic and its toxicity is related to its physicochemical specification. It has played a key role in extracting gold and other metals such as silver, copper and zinc from ores in many countries of the world. Cyanides are highly toxic and their toxicity is related to their physicochemical speciation (Gurbuz et al., 2009). Cyanide treatment by microorganisms for tailings and wastewaters is a proven and viable alternative to chemical processes. For over two decades, microbial treatment of cyanide has been investigated by many researchers (Knowles, 1976; Patil and Paknikar 1999; Adams et al., 2001)

The detoxification of cyanide by algae was examined by exposing cultured suspensions of *Arthrospiramaxima, Chlorella* sp. and *Scenedesmus obliquus* at pH of 10.3. The removal efficiency was 99%. This study explored the use of algae in this field under extreme conditions with high removal efficiency. A novel study has confirmed the use of microalgae in the bioremediation of cyanide containing waste (Gurbuz et al., 2004; 2009). Chapatwala et al (1998) studied the biodegradation of cyanides, cyanates and thiocyanates by *Pseudomonas putida* and found that cyanide compounds were used as sole source of carbon and nitrogen. The degradation of cyanides has been observed in several bacteria including *Pseudomonas stutzeri* AK61, *Alcaligenes xylosoxidans subsp. denitrificans* DF3 and *Bacillus pumilus* C1 Cyanide was catalysed by a cyanide dihydratase. (Ingvorsen et al. 1991; Meyers et al. 1991;

Watanabe et al. 1998). *Klebsiella oxytoca* has shown to have another patway which cyanide was catalysed by nitrogenase (Kao et al., 2003).

4.2. Uses in Detoxification of Phenols

The occurrence of phenols as industrial pollutants in surface waters is quite common due to the release of by-products in the petrochemical industry, coal liquefaction plants, production of plastics and dyes, as well as in the pulp industry (Neilson et al 1991). Some of the phenolic compounds are suspected to be endocrine disrupters and have adverse effects on humans and other organisms in the natural ecosystem at concentrations lower than the emission standard for phenols (Staples, et al., 1998). Therefore the removal of toxic phenolic compounds from industrial wastewater is an important issue to be addressed. Bacteria have been exposed wastewater rich in phenolic compounds including *Pseudomonas* sp. (Chitra et al., 1995), *Rhodococcus* sp. (Oh and Han, 1997), *Phanerochaere* sp. (Perez et al., 1997). When bacteria are used for the removal of phenols from effluent, it is necessary to add an external organic carbon source to maintain biomass and the ability to remove these compounds. Hence, microalgae seems to be promising organism as they can grow photoautotrophitally without the addition of an organic carbon source. *Chlorella fusca* var. *vacuolata* and *Anabaena variabilis* were showed high 2,4-dinitrophenol (DNP) removal ability over the concentration range of 5 to 40 pM (Hirooka, et al., 2003). In the study of Tarlan et al (2002) the ability of algae to treat a wood-based pulp and paper industry wastewater was investigated. Algae were found to remove up to 58% of COD, 84% of color and 80% of AOX (absorbable organic halides) from pulp and paper industry wastewaters under different lighting and initial wastewater strength conditions. They have found no remarkable differences were observed in COD and color removal when light intensity and wastewater strength were changed, while AOX removals were strongly affected. Pinto et al (2002) showed that two green microalgae, *Ankistrodesmus braunii* and *Scenedesmus quadricauda*, degraded phenols (each tested at 400 mg ml−1) selected from olive-oil mill wastewaters, within 5 days, with a removal greater than 70%. The presence of a high phenol concentration in the medium did affect algal morphology, during the study some algal morphology. During the study some algal cells became larger, many oil droplets were observed in the cytoplasm and some cells showed anomalous shapes. Further studies on algal removal of phenols were pursued by Pinto et al. (2003) and they found that higher removal of phenols was obtained in the dark, for both strains. Nakai et al. (2001) previously shown that the toxicity of simple phenols is mainly due to autoxidation processes which are enhanced by light. Another hypothesis has been suggested by Chandra (2001). Author has found that microbial decolorization of pulp and paper mill effluent was influenced by the levels of nitrogen and phosphorus. Lima et al. (2003) reported p-nitrophenol removal of 50mgl^{-1} d^{-1} by a consortium of *Chlorella vulgaris* and *Chlorella pyrenoidosa* under non-optimized conditions, which was close to the 100 mgl^{-1} d^{-1} achieved with *Pseudomonas sp.* by Kulkarni and Chaudhari (2006). Chaillan et al. (2006) described the appearance of cyanobacterial mats (*Phormidium animale*) in petroleum-polluted site. Although cyanobacterium found to have no direct effect on biodegradation of crude oil, the degradation activity being exclusively achieved by the other microorganisms present in the microbial consortium of the mat.

In recent years an enzyme-catalyzed polymerization and precipitation process has been explored as a new method for the treatment of aqueous phenols (Wu et al., 1998), Several researchers proposed that laccases might be useful for removing phenolic contaminants from water or wastewater (Thurston, 1994). This approach has also been taken to remove bisphenol-A (BPA) (Hibrano et al., 2000; Tsutsumi et al., 2001). BPA is suspected to be an endocrine disrupting chemical (EDC) for hormonal signals in animals and has an irreversible effect on the development of the reproductive organs (Laws et al., 2000) and has been heavily released to surface waters and lands (US EPA TRI, 2004). Modaressi et al. (2005) have been presented that using the enzyme laccase to treat synthetic wastewater containing bisphenol-A was success by the presence of PEG(polyethylene glycol) enabled >95% removal of BPA.

Enzyme-based biosensors represent potential alternatives to the analysis or screening of phenolic compounds in environmental samples and in addition as alarm systems. The most commonly employed biosensor approach for phenol analysis is the combination of the reaction catalysed by a phenol oxidase, e.g. tyrosinase and laccase, with a variety of detection principles (Burestedt et al 1996; Hedenmo et al 1997). Two amperometric biosensors based on the enzymes cellobiose dehydrogenase (CDH) and quinoprotein-dependent glucose dehydrogenase (GDH), have been applied for monitoring the phenolic content in water samples, collected at different stages of a waste water treatment process (Nistor et al., 2000)

5. Notes on the Resistance to Heavy Metals

The mechanisms involved in bacterial metal resistance result from either the active efflux pumping of the toxic metal out of the cell, or the enzymatic detoxification converting a toxic ion into a less tôxic or less available metal ion. The genes encoding these resistance systems are most often located on plasmids or transposons. (Silver 1992, 1996). A range of specific and nonspecific metal-binding compounds are produced by microorganisms. Nonspecific metal-binding compounds range from simple organic acids and alcohols to macromolecules, such as polysaccharides, proteins, humic and fulvic acids (Birch and Bachofen, 1990; Bridge et al., 1999; Sayer and Gadd, 2001). These extracellular substances (EPS) are produced by bacteria, algae and fungi and also bind potentially toxic metals (White and Gadd, 1998).

Microalgae employ a variety of biochemical strategies to reduce the toxicity of non-essential trace metals (e.g., Hg and Cd) and to metal homeostasis in the cytoplasm (Hall, 2002; Perales-Vela et al., 2006). Research in microbial cells has shown that organisms tend to utilize only one of the mechanisms for basic heavy metal tolerance (Rigle and Kumamoto, 2000). Cobbett and Goldsbrough, (2002) reported that microalgae and some fungi have preferentially developed the production of peptides capable to bind heavy metals. These molecules, as organometallic complexes, are further partitioned inside vacuoles to facilitate appropriate control of the cytoplasmic concentration of heavy metal ions, thus preventing or neutralizing their potential toxic effect. *Pseudomonas fluorescens* was studied to elucidate the differential protein expression of this bacterium when exposed to heavy metal stress, using two-dimensional electrophoresis. The proteins spo VG and enolase showed upregulation during the bacterial exposure to lead and copper (Sharma et al., 2006).

Metallothioneins (MTs) are a class of ubiquitously occurring low molecular weight cysteine- and metal-rich proteins containing sulfur-based metal clusters such as class II metallothioneins (MTII) (Romero-Isart and Vasak, 2002; Vasak, 2005) and non-translationally synthesized polypeptides some times described as class III metallothioneins (phytochelatins— PCs or MTIII) (Grill et al., 1985; Hirata et al., 2001; Mendoza-Co´zatl et al., 2005; Perales-Vela et al., 2006). These proteins are found in animals, higher plants, eukaryotic microorganisms and some prokaryotes (Howe et al., 1997) and they are products of mRNA translation and its enzymatic synthesis (Romero-Isart and Vasak, 2002; Cobbett and Goldsbrough, 2002). The prokaryotic MTs identified and are found in a few cyanobacterial strains of the genus *Synechococcus*. These MTs, which are encoded by the *smtA* gene, contains fewer cysteine residues than mammalian MTs (Ji and Silver 1995). A CPx-ATPase (ZiaA) which handles zinc tolerance by effluxing excess zinc out of the cytosol has been identified in *Synechocystis* PCC 6803 (Thelwell et al 1998). Liu et al (2003) have identified and characterized A metallothionein (BmtA) and a CPx-ATPase (Bxa1) from the cyanobacterium *Oscillatoria brevis*.

Microalgae employed two other main proteins other than MTs, on the detoxification of hazardous heavy metal ions. The amino acid cysteine (Cys) contains a sulfhydryl (thiol group, –SH) which is the site of metal binding. Cys-containing peptides such as glutathione (GSH) are responsible for metal sequestration in living cells. A Cd-tolerant strain of the alga *Chlamydomonas reinhardtii* has significantly higher levels of Cys than the Cd-sensitive strain (Hu et al., 2001). A Cys pool may thus be critical to algal defense against metal toxicity. In addition to cysteine, proline (Pro) is another group of protein that has been shown to play an important role in heavy metal stress in some but not all algal species. Transgenic microalgae *C. reinhardtii* expressing a mothbean D1-pyrroline-5- carboxylate synthetase (P5CS) gene showed higher free-Pro levels than wild-type cells, grew to higher densities than wild-type cells in the presence of toxic concentrations of Cd, and sequestered four times more Cd per cell than wild-type cultures (Siripornadulsil et al., 2002). Accumulations of this amino acid may permit osmotic adjustment and provide protection for enzymes, biological membranes and polyribosomes by forming stable complexes with free radicals that could otherwise prove toxic.

The Metalloregulatory (MerR) proteins are a group of transcriptional factors that control metal ion, radical, and small organic molecule concentrations inside bacterial cells. The MerR proteins are present in most bacterial genomes and these proteins typically regulate defensive systems against toxic or elevated levels of metal ions. Members from this protein family have been found to sense and control levels of Cd^{2+}, Zn^{2+}, Co^{2+}, Cu^{+} , Ag^{+} , Au^{+} , Hg^{2+}, and Pb^{2+} ions inside various bacteria (Brown et al., 2003). These proteins are categorized according to their affinity towards specific metals. CadR, CoaR, and GolS are MerR family proteins that have been shown to selectively respond to Cd(II), Co(II), and Au(I), respectively Chen and He 2008). ArsR exhibits selective recognition of As^{3+} and Sb^{3+} ions, meanwhile, other members of this protein recognize toxic heavy metals (such as Cd^{2+}, Pb^{2+}, Bi^{3+}) and biologically essential metal ions (such as Co^{2+}, Ni^{2+}, Zn^{2+}) (Busenlehner et al 2003). PbrR is found in *Ralstonia metallidurans* CH_3^4, a bacterium specifically adapted to survive under toxic heavy metal environment. The PbrR protein is responsible for regulation of lead (II) efflux pumps involved in lead detoxification inside *R. metallidurans* (Borremans et al 2001). MntR exhibits a high selectivity for Mn^{2+} over Fe^{2+} (Kehres and Maguire 2003).

There are other mechanisms that microorganisms become resistant to surrounding habit when it contains metals. Methylation is another way of bacteria and fungi can employ under aerobic and anaerobic conditions for Hg, As, Se, Sn, Te and Pb removal. Methyl groups are enzymatically transferred to the metal, and a given species may transform a number of different metals. In this case methylated metals differ in their solubility, volatility and toxicity (Gadd 1993)

6. Conclusion

The potential application of biological and enzymatic treatments of various toxic liquid wastes has advantages and disadvantages. Bioprocesses may offer advantages like, milder reaction conditions (temperature, pressure, and pH); less hazardous operation and reduced environmental impact; greater specificity of catalytic reaction; less expensive or more readily available raw materials., improved process efficiencies (e.g., higher yields, reduced energy consumption). Other than above advantages rDNA technology may be in use to develop new processes. On the other hand, there are some limitations of the bioprocesses, such as, generation of complex product mixtures requiring extensive separation and purification, contamination by foreign organisms; the necessity to provide and handle large volumes of process water and to dispose of equivalent volumes of high biological oxygen demand wastes; establishing suitable environment and in addition to this, providing the nutritional needs of living cells.

Nevertheless, due to the increasing need for the safe removal of toxic metals and the toxicants and the continued public interest to environmental problems, the selection of organisms which are resistant to toxicants has been playing key role in environmental control programs. Commercial utilization and the development of these bioprocess systems and the use of bicatalysts is desirable on both environmental and economic aspects.

References

Aitken, MD. Waste treatment applications of enzymes opportunities and obstacles. *The Chemical Engineering Journal.* 1993, 52, B49-B58.

Ackerley, DF; Gonzalez, CF; Park, CH; Blake, R; Keyhan, M; Matin, A. Chromate-Reducing Properties of Soluble Flavoproteins from *Pseudomonas putida* and *Escherichia coli. Applied And Environmental Microbiology/* 2004, 70, 873–882

d'Acunzo, F; Galli, C; Masci, B. Oxidation of phenols by laccasemediator systems. *Eur. J. Biochemistry.* 2002, 269(12), 5330–5335.

Adams, DJ; Komen, JV; Pickett, TM; Biological cyanide degradation. In: Young, C. (Ed.), Cyanide: Social, Industrial and Economic Aspects. The Metal Society, United States, 2001, 203–213.

Allsopp D; Seal K; Gaylarde C. Introduction to biodeterioration. Cambridge University Press; 2004, ISBN 0521528879 , 1–256.

Arica, MY; Bayramoğlu, G. Cr(VI) biosorption from aqueous solutions using free and immobilized biomass of *Lentinus sajor-caju*: Preparation and kinetic characterization. Colloids and Surfaces A: *Physicochemical Engineering Aspects.* 2005, 253, 203–211.

Ayotte, JD; Montgomery, DL; Flanagan, SM; Robinson, KW. Arsenic in groundwater in eastern New England: occurrence, controls, and human health implications. *Environmental Science and Technology.* 2003, 372075–372083.

Balaban, N; Gabdrakhmanova, L; Sharipova, M; Sokolova, E; Malikova, L; Mardanova, A; et al. Selection of cultivation medium for production of late stationary phase serine proteinases from *Bacillus intermedius*. *J. Basic Microbiology.* 2004, 44, 415–423.

Basheer, S; Kut, ÖM; Prenosil, JE; Bourne, JR. Development of an enzyme membrane reactor for treatment of cyanide-containing wastewaters from the food industry. *Biotechnology and Bioengineering.* 1993, 41, 465-473.

Berno, E; Pereira Marcondes, DF; Gamalero, SR; Eandi, M. Recombinant Escherichia coli for the biomonitoring of benzene and its derivatives in the air Ecotoxicol. *Environmental Safety.* 2004, 57, 118–122.

Beveridge, TJ; Doyle, RJ. *Metal Ions and Bacteria.* 1989, Wiley, New York.

Biran, I; Rissin, DM; Ron, EZ; Walt, DR. Optical imaging fiber-based live bacterial cell array biosensor. *Anal Biochemistry.* 2003, 315, 106–113.

Birch, L., Bachofen, R. Complexing agents from microorganisms. *Experientia.* 1990, 46, 827–834.

Bishnoi, N; Kumar, R; Kumar, S; Rani, S. Biosorption of Cr(III) from aqueous solution using algal biomass S*pirogyra* spp, *J. Hazardous Material.* 2007, 145, 142–147.

Bridge, TAM; White, C; Gadd, GM. Extracellular metalbinding activity of the sulphate-reducing bacterium. *Desulfococcus multivorans. Microbiology.* 1999, 145, 2987–2995.

Bocchini, DA; Damiano, VB; Gomes, E; Da Silva, R. Effect of *Bacillus circulans* D1 thermostable xylanase on biobleaching of eucalyptus kraft pulp. *Appl. Biochem. Biotechnology.* 2003, 106(13), 393–402.

Bogan, B; Lamar, RT. Polycyclic aromatic hydrocarbon-degrading capabilities of *Phanerochaete laevis* HHB-1625 and its extracellular ligninolytic enzymes. *Appl. Environ. Microbiology.* 1996, 62,1597– 1603.

Borremans B; Hobman JL; Provoost A; Brown NL; Van der Lelie, D. Cloning and functional analysis of the pbr lead resistance determinant of Ralstonia metallidurans CH_3^4. *J. Bacteriol.* 2001, 183, 5651-5658.

Boucabeille, C; Bories, A; Ollivier, P. Degradation of thiocyanate by a bacterial coculture, *Biotechnology Letters.* 1994, 16, 425–430.

Burestedt, E; Narvaez, A; Ruzgas, T; Gorton, L; Emneus, J; Dominguez, E; Marko-Varga, G. *Anal. Chem.* 1996, 68, 1605.

Busenlehner, LS; Pennella, MA; Giedroc, DP: The SmtB/ArsR family of metalloregulatory transcriptional repressors: structural insights into prokaryotic metal resistance. *FEMS Microbiol Review.* 2003, 27, 131-143.

Brown, NL; Stoyanov, JV; Kidd, SP; Hobman, JL. The MerR family of transcriptional regulators. *FEMS Microbiol Review.* 2003, 27,145-163.

Cai, J., Salmon, K., DuBow, M.S., 1998. A chromosomal *ars* operon homologue of *Pseudomonas aeruginosa* confers increased resistance to arsenic and antimony in *Escherichia coli. Microbiology.* 144, 2705–2713.

Cain, A; Vannela, R; Woo, LK. Cyanobacteria as a biosorbent for mercuric ion *Bioresource Technology.* 2008, 99, 6578–6586.

Camargo, FAO; Bento, FM; Okeke, BC; Frankenberger, WT. Chromate reduction by chromium resistant bacteria isolated from soils contaminated with dichromate. *J. Environmental Quality.* 2003, 32, 1228–1233.

Campanella, L; Cubadda, P; Sammartino, MP; Saoncella, A. *Water Research.* 2000, 35, 69–76.

Cervantes, C; Guangyong, J; Ramirez, JL; Silver, S. Resistance to arsenic compounds in microorganisms. *FEMS Microbiol. Review.* 1994, 15, 355–367.

Chaillan, F; Gugger, M; Saliot, A; Coute′, A; Oudot, J. Role of cyanobacteria in the biodegradation of crude oil by a tropical cyanobacterial mat. *Chemosphere.* 2006, 62, 1574–1582.

Chandra, R. Microbial decolorization of pulp and paper mill effluent in the presence of nitrogen and phosphorus by activated sludge process. *Journal of Environmental Biology.* 2001, 22, 23–27.

Chapatwala, KD; Babu, GRV; Vijaya, OK; Kumar, KP; Wolfram, JH. Biodegradation of cyanides, cyanates and thiocyanates to ammonia and carbon dioxide by immobilized cells of *Pseudomonas putida, J. Ind. Microbiol*ogy. 1998, 20 28–33.

Chen, PR; He, C. Selective recognition of metal ions by metalloregulatory proteins. *Current Opinion in Chemical Biology.* 2008, 12, 214–221.

Chitra, S; Sekaran, G; Padmavathi, S; Chandrakasan, G. Removal of phenolic compounds from wastewater using mutant strain of *Pseudomonas pictorum. J. Gen. Appl. Microbiol.* 1995, 41, 229–237.

Cho, D., Lee, S., Park, S. and Chung, A., Studies on the biosorption of heavy metals onto *Chlorella vulgaris. Journal of Environmental Science and Health A,* 1994, *29*,389–409.

Cho, J-C; Park, K-J; Ihm, H-S; Park, J-E; Kim, S-Y; Kang,I., Lee, K.-H., Jahng, D. Lee,D.-H. Kim, S.-J., A novel continuous toxicity test system using a luminously modified freshwater bacterium. *Biosens. Bioeletronics.* 2004, 20338–344.

Cho, DH; Kim EY. Characterization of Pb^{2+} biosorption from aqueous solution by *Rhodotorula glutinis, Bioprocess Biosyst. Eng.* 2003, 25, 271–277.

Chouteau, C; Dzyadevych, S; Chovelon, JM; Durrieu, C. Development of novel conductometric biosensors based on immobilized whole cell *Chlorella vulgaris* microalgae. *Biosensor Bioelectronics.* 2004, 19, 1089–1096.

Clark, JM; Switzer, RL. Experimental Biochemistry. Second Edition, W.H. Freeman and Company, NY, USA. 1977.

Cobbett, C; Goldsbrough, P. Phytochelatin and metallothioneins: Roles in heavy metal detoxification and homeostasis. *Annu. Rev. Plant Biology.* 2002, 53, 159–182.

Cooper, VA; Nicell, JA. Removal of phenols from a foundry wastewater using horseradish peroxidase. *Water Research.* 1996, 30(4), 954-964.

Coppella, SJ; DelaCruz, N; Payne, GF; Pogell, BM; Speedie, MK; Karns, JS; Sybert, EM; Connor, MA. Genetic engineering approach to toxic waste management:case study for organophosphate waste treatment. *Biotechnology Progress.* 1990, 6, 76-81.

Cullen, W; Reimer, K. Arsenic speciation in the environment. *Chemical Reviews.* 1989, 89, 713–764.

Deng X; Wilson, DB. Bioaccumulation of mercury from wastewater by genetically engineered *E. coli, Appl. Microbiol. Biotechnology.* 2001, 56 276–279.

Deng, X; Yi, XE; Liu, G. Cadmium removal from aqueous solution by gene-modified *Escherichia coli* JM109. *Journal of Hazardous Materials.* B139, 2007, 340–344.

Daulton, TL; Little, BJ; Jones-Meehan J; Blom DA; Allard LF. Microbial reduction of chromium from the hexavalent to divalent state *Geochimica et Cosmochimica Acta.* 2007, 71556–71565

Duff, SJ; Moritz, JW; Casavant, TE. Effect of surfactant and particle size reduction on hydrolysis of deinking sludge and nonrecyclable newsprint. *Biotechnology and Bioengineering.* 1995, 45, 239-244.

Dursun AY; Uslu, G; Tepe, O; Cuci, Y; Ekiz, HI. A comparative investigation on the bioaccumulation of heavy metal ions by growing *Rhizopus arrhizus* and *Aspergillus niger*, *Biochem. Eng. Journal.* 2003, 15, 87–92.

Eriksson KE. Biotechnology in the pulp and paper industry. *Adv. Biochem. Eng. Biotechnol.* 57, 1997, 339.

Erhan, E; Keskinler, B; Akay, G; Algur, OF. Removal of phenol from water by membrane-immobilized enzymes. Part I. Dead-end filtration. *Journal of Membrane Science.* 2002, 206, 361–373.

Farré, M; Brix,R; Barceló D. Screening water for pollutants using biological techniques under European Union funding during the last 10 years. *Trends in Analytical Chemistry.* 2005, 24, 532-545.

de Felice, B; Pontecorvo, G; Carfagna, M. Degradation of wastewaters from olive oil mills by Yarrowia lipolytica ATCC 20255 and *Pseudomonas putida. Acta Biotechnology.* 2004, 17, 231–239.

Ferrer, I; Dezotti, M; Duran, N. Decolorization of kraft effluent by free and immobilized lignin peroxidases and horseradish peroxidase. *Biotechnology Letters.* 1991, 13, 577-582.

Freire, RS; Duran, N; Kubota, LT. An Optical Biosensor based on Immobilization of Laccase and MBTH in Stacked Films for the Detection of Catechol. *Anal. Chim. Acta.* 2002, 463, 229–238.

Gabr, RM; Hassan, SHA; Shoreit, AAM.Biosorption of lead and nickel by living and non-living cells of *Pseudomonas aeruginosa* ASU 6a. *International Biodeterioration & Biodegradation.* 2008, 62, 195–203

Gadd, GM. Microbial formation and transformation of organometallic and organometalloid compounds. *FEMS Microbiol. Review.* 1993, 11, 297–316.

Gadd, GM. Heavy metal pollutants: environmental and biotechnological aspects. In: Lederberg J, editor. *The encyclopedia of microbiology.* San Diego: Academic Press Inc; 2000, 607–17.

Gailer, J. Arsenic–selenium and mercury–selenium bonds in biology Coordination *Chemistry Reviews.* 2007, 251, 234–254.

Garnham, G.W., Green, M., Chromate (VI) uptake by and interaction with cyanobacteria. *Journal of Industrial microbiology.* 1995, 247–251.

Gavrilescu, M. Removal of heavy metals from the environment by biosorption, *Life Sci. Eng.* 2004, 4, 219–232.

Gurbuz, F; Ciftci, H; Akcil, A. Karahan A.G., Microbial detoxification of cyanide solutions: a new biotechnological approach using algae, *Hydrometallurgy.* 2004, 72, 167–176.

Gurbuz F., Ciftci, H., Akcil, A., Biodegradation of cyanide containing effluents by *Scenedesmus obliquus*. *Journal of Hazardous Materials.* 2009, 162, 74-79.

Golab, Z. Breitenbach M. Sites of copper binding in *Streptomyces pilosus.Water Air Soil Pollution.* 1995, 82, 713–21.

Hall, JL. Cellular mechanism for heavy metal detoxification and tolerance. *J. Experimental Botanics*. 2002, 366, 1–11.

Hasan, F; Shah, AA; Hameed, A. Industrial applications of microbial lipases, *Enzyme and Microbial Technology*. 2005, 39, 235–251.

Hattunen, T; Finell, M; Salminen, S. Arsenic removal by native and chemically modified lactic acid bacteria. *International Journal of Food Microbiology*. 2007, 120, 173–178.

Hedenmo, M; Narváez, A; Domínguez E., Katakis I., Improved mediated tyrosinase amperometric enzyme electrodes, *J. Electroanal. Chemistry*. 1997, 425, 1.

Hirata, K; Tsujimoto, Y; Namba, T; Ohta, T; Hirayanagi, N; Myiasaka, H; Zenk, MH; Myamoto, K. Strong induction of phytochelatin synthesis by zinc in marine green algae, (*Dunaliela tertiolecta*). *J. Biosci. Bioengineering*. 2001, 92, 24–29.

Hibrano, T; Hondo, Y; Watanabe, T; Kuwahara, M. Degradation of bisphenol-A by the lignin-degrading enzyme, manganese peroxidase, produced by white-rot basidiomycete, *Pleurotus Ostreatus*. *Biosci. Biotechnol. Biochem*. 2000, 64, 1958–1962.

Hirooka T; Akiyama Y; Tsuji N; Nakamura T; Nagase, H; Hirata K; Miyamoto K. Removal of hazardous phenols by microalgae under photoautotrophic conditions. *Journal of Bioscience and Bioencineering*. 2003, 95, 200-203.

Howe, R; Evans, RL; Ketteridge, SW. Copper-binding proteins in ectomycorrhizal fungi. *New Phytol*. 1997, 135, 123–131.

Hu, S; Lau, KWK; Wu, M. Cadmium sequestration in *Chlamydomonas reinhardtii*. Issue Series Title: *Plant Sciences*. 2001, 161, 987–996.

Ingvorsen, K; Hojerpedersen, B; Godtfredsen, SE. Novel cyanide-hydrolysing enzyme from *Alcaligenes xylosoxidans* subsp. denitrificans. *Appl. Environ. Microbiology*. 1991, 57, 1783 –1789.

Jandhyala, D; Berman, M; Meyers, PR; Sewell, BT; Willson, RC; Benedik, MJ. CynD, the cyanide dihydratase from *Bacillus pumilus*: gene cloning and structural studies. *Appl. Environ. Microbiology*. 2003, 69, 4794-4805.

Jaspers, C; Jimenez, G; Penninkx, M. Evidence for a role of Mn peroxidase in the decolorization of Kraft pulp bleach plant effluent by *P. chrysosporium*: effects of initial culture conditions on enzyme production. *Journal of Biotechnology*. 1994, 37, 229–234.

Ji, G; Silver, S. Bacterial resistance mechanisms for heavy metals of environmental concern. *J. Ind. Microbiol*. 1995, 14, 61–75.

Jördening H-J, Winter J, Environmental biotechnology: concepts and applications. Weinheim7 Wiley- VCH; 2004.

Kang, S-Y; Lee, J-U; Kim, K-W. Biosorption of Cr(III) and Cr(VI) onto the cell surface of *Pseudomonas aeruginosa*. *Biochem. Eng. Journal*. 2007, 36, 54–8.

Kansoh, AL; Nagieb, ZA. Xylanase and Mannanase enzymes from *Streptomyces galbus* NR and their use in biobleaching of softwood kraft pulp. *Antonie van Leeuwenhoek*. 2004, 85(2), 103–114.

Kao, CM; Liu, JK; Lou, HR; Lin, CS; Chen, SC. Biotransformation of cyanide to methane and ammonia by *Klebsiella oxytoca*. *Chemosphere*. 2003, 50, 1055 –1061.

Karam, J; Nicell, JA. Potenital applications of enzymes in waste treatment. *J. Chem. Tech. Biotechnology*. 1997, 69, 141-153.

Katırcıoğlu H; Aslım, B; Rehber Türker, A; Atıcı, T; Beyatlı, Y. Removal of cadmium(II) ion from aqueous system by dry biomass, immobilized live and heat-inactivated *Oscillatoria* sp. H1 isolated from freshwater (Mogan Lake) *Bioresource Technology*. 2008, 99, 4185–4191

Kehres, DG; Maguire, ME. Emerging themes in manganese transport, biochemistry and pathogenesis in bacteria. *FEMS Microbiol Review*. 2003, 27, 263-290.

Kerper, LE; Ballatori, N; Clarkson, TW. Methyl mercury transport across the blood-brain barrier by an amino acid carrier. *Am. J. Physiology*. 1992, 262, 761–765.

Klibanov, AM; Alberti, BN; Morris, ED; Felshin, ZM. Enzymatic removal of toxic phenols and anilines from wastewaters. *J. Appl. Biochem.* 1980, 2, 414–421.

Knowles, CJ;Microorganisms and cyanide. *Bacteriological Review*. 1976. 40, 652–680.

Kostal, J; Yang, R; Wu, CH. Mulchandani, A., Chen, W., Enhanced arsenic accumulation in engineered bacterial cells expressing ArsR. *Appl. Environmental Microbiology*. 2004, 70(8), 4582–4587.

Kulkarni, M; Chaudhari, A. Biodegradation of p-nitrophenol by *P. putida. Bioresource Technology*. 2006, 97, 982–988.

Lai, YC; Lin, SC. Application of immobilized horseradish peroxidase for the removal of *p*-chlorophenol from aqueous solution, *Process Biochem*. 2005, 40, 1167–1174.

Lanciotti, R; Gianotti, A; Baldi, D; Angrisani, R; Suzzi, G; Mastrocola, D; Guerzoni, M.E. Use of Yarrowia lipolytica strains for the treatment of olive mill wastewater. *Bioresource Technology*. 2005, 96, 317–322.

Lanyon, YH; Marrazza, G; Tothhill, IE; Mascini, M. Benzene analysis in workplace air using an FIA-based bacterial biosensor. *Biosens. Bioeletron*ics. 2005, 20 2089–2096.

Laws, SC; Carey, SA; Ferrell, JM; Bodman, GJ; Cooper, RL. Estrogenic activity of octylphenol, nonylphenol, bisphenol A and methoxychlor in rats. *Toxicol. Sciences*. 2000, 54, 154–167.

Lima, SAC; Castro, PML; Morais, R. Biodegradation of p-nitrophenol by microalgae. *J. Appl. Phycology*. 2003, 15, 137–142.

Li, ZJ; Yuan, HL. Characterization of cadmium removal by *Rhodotorula* sp. Y11, *Appl. Microbiol. Biotechnol.* 2006, **73**, 458–463.

Liu, T; Nakashima, S; Hirose, K; Uemura, Y; Shibasaka, M; Katsuhara, M; Kasamo, KA. metallothionein and CPx-ATPase handle heavy-metal tolerance in the Flamentous cyanobacterium *Oscillatoria brevis FEBS Letters*. 2003, 542, 159-163

Lukavsky, J; Maršálek, B. The evaluation of toxicity by a biosensor with immobilized algae. *Algol. Stud.* 1997, 85, 147–155.

Lopez, A; Larao N; Priergo, J; Marques, A. Effect of pH on the biosorption of nickel and other heavy metals by *Pseudomonas fluorescens* 4F39. *Journal of Industrial Microbiology and Biotechnology*. 2001, 24, 146–151.

Maeda, S. Biotransformation of arsenic in the freshwater environment. In: Nriagu JO. (Ed.) *Arsenic in the Environment. Advances in Environmental Science and Technology*. Wiley:New York, 1994, 155–187.

Macaskie, LE; Wates, JM; Dean, ACR. Cadmium accumulation by a *Citrobacter* sp. immobilized on gel and solid supports: applicability to the treatment of liquid wastes containing heavy metal cations. *Biotechnology and Bioengineering*. 1987, 30, 66-73.

Majcherczyk, A; Johannes, C; Huttermann, A. Oxidation of polycyclic aromatic hydrocarbons (PAH) by laccase of *Trametes versicolor*. *Enzyme Microbiol Technologyl*. 1998, 22, 335–341.

Malik, A. Biotechnological potential of immobilized algae for wastewater N, P and metal removal: A review, *BioMetals*. 2002, 15, 377-390.

Malik, A. Metal bioremediation through growing cells. *Environ. International*. 2004, 30, 261–278.

Marqués, AM; Roca, X; Simon-Pujol, MD; Fuste MC; Congregado, F. Uranium accumulation by *Pseudomonas sp.* EPS-5028. *Applied Microbiology and Biotechnology*. 1991, 35, 406–410.

Marten, E; Muller, RJ; Deckwer, WD. Studies on the enzymatic hydrolysis of polyesters. I. Low molecular mass model esters and aliphatic polyesters. *Polym. Degrad. Stab*. 2003, 80, 485–501.

Mawadza, C; Hatti-Kaul, R; Zvauya, R; Mattiasson, B. Purification and characterization of cellulases produced by two *Bacillus* strains. *J. Biotechnology*. 2000, 83,177–87.

McLean, JS; Lee, JU; Beveridge, TJ. Interactions of bacteria and environmental metals, fine-grained mineral development, and bioremediation strategies. In: Huang, P.M., Bollag, J.-M., Senesi, N. (Eds.), *Interactions between Soil Particles and Microorganisms*. Wiley, New York, 2002, 227–261.

Meinrath, G; Kato, Y; Kimura, T; Yoshida, Z. Solid-aqueous phase equilibria of uranium(VI) under ambient conditions. *Radiochimica Acta*. 1996, 75, 159–167.

Mendoza-Cózatl, D; Losa-Tavera, H; Hernández-Navarro, A; Moreno-Sánchez, R. Sulfur assimilation and glutathione metabolism under cadmium stress in yeast, protists and plants. *FEMS Microbioogy. Review*. 2005, 29, 653–671.

Meyers, PR; Gokool, P; Rawlings, DE; Woods, DR. An efficient cyanide-degrading *Bacillus pumilus* strain. *J. Gen. Microbiology*. 1991, 137, 1397 –1400.

Michael, FC; Dass, SB; Grulke, EA; Reddy, CA. Role of LiP and MnP of *Phanerochaete chrysosporium* in the decolorization of Kraft bleach plant effluent. *Applied and Environmental Microbiology*. 1991, 57, 2368–2375.

Modaressi, K; Taylor, KE; Biswas, N. Laccase-catalyzed removal of bisphenol-A from water: Protective effect of PEG on enzyme activity. *Water Research*. 2005, 39, 4309–4316.

Moreno-Garrido, I. Microalgae immobilization: Current techniques and uses. *Bioresource Technology*. 2008, 99, 3949–3964.

Moore, JW; Ramamoorthy, S. Heavy Metals in Natural Waters. Springer-Verlag, New York, pp. 1984, 28–57, 77–99, 182–184.

Moreira dos Santos, M; Soares, AMVM; Ribeiro, R. An in situ bioassay for freshwater environments with the microalga Pseudokirchneriella subcapitata. *Ecotox. Environ. Safety*. 2004, 59, 164–173.

Moreira, SM; Moreira-Santos, M; Guilhermino, L; Ribeiro, R. Immobilization of the marine microalga Phaeodactylum tricornutum in alginate for in situ experiments: bead stability and suitability. *Enzyme Microbiolgy Technology*. 2006, 38, 135–141.

Moriya, RY; Goncalves, AR; Duarte, MC. Sugarcane bagasse pulps: biobleaching with commercial cartazyme HS and with *Bacillus pumilus* xylanase. *Appl. Biochem. Biotechnol*. 2005, 121–124,171–81.

Munôz, R; Guieysse, B. Algal–bacterial processes for the treatment of hazardous contaminants: a review. *Water Resources*. 2006, 40, 2799–2815.

Munoz, R; Alvarez, MT; Munoz, A; Terrazas, E; Guieysse, B; Mottiasson, B. Sequential removal of heavy metals ions and organic pollutants using an algal bacterial consortium. *Chemosphere*. 2006, 63, 893–1062.

Murray, LA; Raab, A; Marr, IL; Feldmann, J. Biotransformation of arsenate to arsenosugars by *Chlorella vulgaris* Appl. *Organometal. Chem.* 2003, 17, 669–674.

Myers, C.R; Carstens, BP; Antholine, WE; Myers, JM. Chromium(VI) reductase activity is associated with the cytoplasmic membrane of anaerobically grown *Shewanella putrefaciens* MR-1. *J. Applied. Microbiology.* 2000, 88, 98–106.

Naessens, M; Leclerc, JC; Tran-Minh, C.. Fiber optic sensor using Chlorella vulgaris for determination of toxic compounds. E*cotoxicology. Environ. Safety.* 2000, 46, 181–185.

Nakai, S; Yutaka, I; Masaaki, H. Algal growth inhibition effects and inducement modes by plant-producing phenols. *Water Res*earch. 2001, 35, 1855–1859.

Neilson, AH; Ellard, AS; Hynning, PA; Ramberger, M. Toxicol. Environ. Chem. (1991)3.

Nicell, JA. Kinetics of horseradish peroxidase-catalysed polymerization and precipitation of aqueous 4-chlorophenol. *Journal of Chemical Technology and Biotechnology.* 1994, 60, 203-215.

Nistor, C; Rose A; Farré, M; Stoica, L; Wollenberger, U; Ruzgas, T; Pfeiffer, D; Barceló, D; Gorton, L; Emnéus, J. In-field monitoring of cleaning efficiency in waste water treatment plants using two phenol-sensitive biosensors *Analytica Chimica Acta.* 2002, 456, 3–17.

Nkhalambayausi-Chirwa, EM; Wang, YT. Simultaneous chromium (VI) reduction and phenol degradation in a fixed-film coculture bioreactor: reactor performance. *Water Research.* 2001, 35(8), 1921–1932.

Ochieng, A; Odiyo, JO; Mutsago, M. Biological Treatment of Mixed Industrial Wastewaters In A Fluidised Bed Reactor *Journal of Hazardous Materials.* 2003, 96, 79–90.

Oh, JS; Han, YH. Isolation and characterization of phenol degrading *Phodococcus sp.* DGUM 2011. *Kor. J. Appl. Microbiol. Biotechnol.* 1997, 25, 459–463.

Ogunseitan, OA; Yang, S; Ericson, J. Microbial d-aminolevulinate dehydratase as a biosensor of lead bioavailability in contaminated environments *Soil Biol. Biochemistry.* 2000, 32, 1899–1906.

Park, CH; Keyhan, M; Wielinga, B; Fendorf, S; Matin, A. Purification to homogeneity and characterization of a novel *Pseudomonas putida* chromate reductase. *Appl. Environ. Microbiology.* 2000, 66(5), 1788–1795.

Park, D; Lim, SR; Yun, YS; Park, JM. Reliable evidences that the removal mechanism of hexavalent chromium by natural biomaterials is adsorption-coupled reduction. *Chemosphere.* 2007, 70, 298–305.

Patil, YB; Paknikar, KM. Removal and recovery of metal cyanides using a combination of biosorption and biodegradation processes. *Biotechnology Letters.* 1999, 21, 913–919.

Paul, A. Hoskisson and Glyn Hobbs Continuous culture – making a comeback? *Microbiology.* 2005, 151, 3153–3159.

Perales-Vela, HV; Penã-Castro, JN; Cannizares-Villanueva, RO. Heavy metal detoxification in eukaryotic microalgae. *Chemosphere.* 2006, 64, 1–10.

Perez, RR; Benito, GG; Miranda, MP. Chlorophenol degradation by *Phanerochaete chrysosporium. Bioresource Technology.* 1997, 60, 207–213.

Pinto, G; Pollio, A; Previtera, L; Temussi, F. Biodegradation of phenols by microalgae. *Biotechnol. Letters.* 2002, 24, 2047–2051.

Pinto, G; Pollio, A; Previtera, L; Stanzione, M; Temussi, F. Removal of low molecular weight phenols from olive oil mill wastewater using microalgae *Biotechnology Letters.* 2003, 25, 1657–1659.

Posorske, LH. Industrial scale application of enzymes to the fat and oil's industry. *J. Am. Oil Chem. Society.* 1984, 61, 1758–1760.

Prakasam, RS; Sherao-Merrie, J; Sheela, N; Saswath, N; Ramakrishra, SV. Biosorption of Chromium VI by Free and Immobilized R*hizopus arrhizus*. *Environ. Pollution.* 1999, 104, 421-427.

Prakasham, RS; Rao, CS; Rao, RS; Rajesham, S; Sarma, PN. Optimization of alkaline protease production by *Bacillus* sp. using Taguchi methodology. *Appl. Biochem. Biotechnology.* 2005, 120, 133–144.

Prasad, MN; Drej, K; Skawinska, A; Stratka, K. Toxicity of cadmium and copper in *Chlamydomonas reinhardtii* wild-type (WT 2137) and cell wall deficient mutant strain (CW 15). *Bulletin of Environmental Contaminant Toxicology.* 1998, 60, 306–311.

Prasad, BB; Banerjee, S; Lakshami, D. An AlgaSORB column for the quantitative sorption of arsenic(III) from water samples. *Water Quality Research Journal of Canada.* 2006, 42, 190–197.

Preetha, B; Viruthagiri, T. Bioaccumulation of chromium(VI), copper(II) and nickel(II) ions by growing *Rhizopus arrhizus Biochemical Engineering Journal.* 2007, 34, 131–135.

Prieto, MB; Hidalgo, A; Serra, JL; Llama, MJ. Degradation of phenol by Rhodococcus erythropolis UPV-1 immobilized on Biolite_ in a packed-bed reactor. *J. Biotechnology.* 2002, 97, 1–11.

Rashamuse, KJ. The bioaccumulation of platinum (IV) from aqueous solution using sulphate reducing bacteria – role of a hydrogenase enzyme.,Msc Thesis, 2003, Rhodes University, South Africa.

Regine, H; Vieira, SF; Volesky, B. Biosorption: A Solution to Pollution? *International Microbiology.* 2000, 3, 17–24.

Riggle, P.J. and Kumamoto, C.A. (2000) *J. Bacteriol.* 182, 4899-4905.

Rogers, KR. Recent advances in biosensor techniques for environmental monitoring. *Analytica Chimica Acta.* 2006, 568, 222–231.

Romero-Isart, N; Vasak, M. Advances in the structure and chemistry of metallothioneins. J. Inorg. *Biochemistry.* 2002, 88, 388–396.

Rothert, A; Deo, SK; Millner, L; Puckett, LG; Madou, MJ; Daunert, S. Whole-cell-reporter-gene-based biosensing systems on a compact disk micro Xuidics platform. *Anal. Biochemistry.* 2005, 342, 11–19.

Russ, R; Rau, J; Stolz, A. The function of cytoplasmic flavin reductases in the reduction of azo dyes by bacteria. *Appl. Environ. Microbiology.* 2000, 66, 1429–1434.

Salinas, E; Elorza de Orellano, M; Rezza, I; Martinez, I; Marchesvky E; Sanz de Tosetti, M. Removal of cadmium and lead from dilute aqueous solutions by *Rhodotorula rubra, Biores. Technology.* 2000, 72, 107–112

Sangave, PC; Pandit, AB Ultrasound and enzyme assisted biodegradation of distillery wastewater. *Journal of Environmental Management.* 2006, 80, 36–46.

Sayadi, S; Ellouz, R. Roles of LiP and MnP in the decolorization of olive mill wastewaters. *Applied and Environmental Microbiology.* 1995, 61, 1098–1103.

Sayer, JA; Gadd, GM. Binding of cobalt and zinc by organic acids and culture filtrates of Aspergillus niger grown in the absence or presence of insoluble cobalt or zinc phosphate. *Mycol. Research.* 2001,105, 1261–1267.

Semple, KT.. Heterotrophic growth on phenolic mixtures by Ochromonas danica. *Res. Microbiology.* 1998, 149, 65–72.

Shackle, V; Freeman, C; Reynolds, B. Exogenous enzyme supplements to promote treatment efficiency in constructed wetlands. *Science of the Total Environment.* 2006, 361, 18– 24.

Shan, D; Mousty, C; Cosnier, S. Subnanomolar Cyanide Detection at Polyphenol Oxidase/Clay. *Biosensors Anal. Chemistry.* 2004, 76, 178–183.

Sharma, BP; Messing, RA. in *Immobilized Enzymes for Food Processing* (ed. Pitcher, W. H.), CRC Press, Boca Raton, Florida, 1980, pp. 185–209.

Sharma, NM; Kumar, S; Sawhney, SK. A novel method for the immobilization of tyrosinase to enhance stability. *Biotechnoloogy Appl. Biochemistry.* 2003,137, 141–148.

Sharma S, Sundaram, CS; Luthra, PM; Singh, Y; Sirdeshmukh, R; Gade, WN. Role of proteins in resistance mechanism of *Pseudomonas fluorescens* against heavy metal induced stress with proteomics approach. *Journal of Biotechnology*, 2006, 126, 374–382.

Shimizu, T; Kudo, K; Nasu, Y. Anaerobic waste activated sludge digestion: a bioconversion mechanism and kinetic model. *Biotechnol. Bioeng.* 1993, 41, 1082–1091.

Shitanda, I; Takada, K; Sakai, Y; Tatsuma, T. Compact amperometric algal biosensors for the evaluation of water toxicity. *Anal. Chim. Acta.* 2005, 530, 191–197.

Siddique, MH; St. Pierre, CC; Biswas, N; Bewtra, JK; Taylor, KE. Immobilized enzyme catalyzed removal of 4-chlorophenol from aqueous solution. *Water Research.* 1993, 27(5), 883-890.

Silver, S. Bacterial heavy metaldetoxification and resistance systems. In *Biotechnology and Environmental Science: Molecular approaches.* (Mongkolsuk, S. et al., eds), 1992, pp. 109–129, Plenum Press

Silver, S. Bacterial heavy metal resistance: New surprises. *Annu. Rev. Microbiology.* 1996, 50, 753–789

Singh, J; Batra, N; Sobti, RC. Purification and characterization of alkaline cellulase produced by a novel isolate, *Bacillus sphaericus* JS1. *J. Ind. Microbiol. Biotechnol.* 2004, 31, 51–6.

Siripornadulsil, S; Traina, S; Verma, DPS; Sayre, RT. Molecular mechanisms of proline-mediated tolerance to toxic heavy metals in transgenic microalgae. *Plant Cell.* 2002, 14, 2837–2847.

Sivalingam, G; Chattopadhyay, S; Madras, G. Solvent effects on lipase catalyzed biodegradation of poly (epsilon-caprolactone) in solution. *Polym. Degrad. Stab.* 2003, 79, 413–418.

Sultan, S; Hasnain, S. Reduction of toxic hexavalent chromium by *Ochrobactrum intermedium* strain SDCr-5 stimulated by heavy metals Sikander 1, Shahida *Bioresource Technology.* 2007, 98, 340–344.

Suzuki, T; Yamaguchi, T; Ishida, M. Immobilization of Prototheca zopfii in calcium alginate vedas for the degradation of hydrocarbons. *Process Biochemistry.* 1998, 33, 541–546.

Staples, CA; Darn, PB; Klecka, G. M., O'Block, ST; Harris, LR. A review of the environmental fate, effects, and exposures of bisphenol A. *Chemosphere.* 1998, 36, 2149-2173.

Takeuchi M, Kawahata H, Gupta L P., Kita N., Morishita Y., Ono Y., Komai T 2007 Arsenic resistance and removal by marine and non-marine bacteria *Journal of Biotechnology.* 2007, 127, 434–442.

Tampion, J; Tampion, MD. Immobilized Cells: Principles and Applications, Cambridge University Press, Cambridge, 1987.

Tangaromsuk, J; Pokethitiyook, P; Kruatrachue, M; Upatham, ES. Cadmium biosorption by *Sphingomonas paucimobilis* biomass. *Bioresource Technology.* 2002, 85, 103–105.

Tarlan, E; Dilek, FB; Yetis, U. Effectiveness of algae in the treatment of a wood-based pulp and paper industry wastewater. *Bioresource Technology.* 2002, 84, 1–5.

Terry, PA; Stone, W. Biosorption of cadmium and copper contaminated water by *Scenedesmus abundans. Chemosphere,* 2002, 47, 249–255

Thacker, U; Parikh, R; Shouche, Y; Madamwar, D. Reduction of chromate by cell-free extract of Brucella sp. Isolated from Cr(VI) ontaminated sites. *Bioresource Technology.* 2007, 98, 1541–1547.

Thelwell, C; Robinson, NJ; Turner-Cavet, JS. *Proc. Natl. Acad. Sci. USA,* 1998, 95, 10728-10733.

Thurston, CF. The structure and function of fungal laccases. *Microbiology.* 1994, 140, 19–26.

Tsezos, M; Volesky, B. The Mechanism of Uranium Biosorption by *Rhizopus arrhizus. Biotechnol. Bioengineering.* 1982, 24, 955-969.

Tsutsumi, Y., Haneda, T., Nishida, T., 2001. Removal of estrogenic activities of bisphenol-A and nonylphenol by oxidative enzymes from lignin-degrading basidomycetes. *Chemosphere.* 42, 271–276.

US Environmental Protection Agency. Toxics Release Inventory (TRI). Data for 2002–2004. Washington, DC 2004.

van Wyk, JPH; Mohulatsi, M. Biodegradation of wastepaper by cellulase from *Trichoderma viride. Bioresource Technology.* 2003, 86, 21–23.

Vasak, M. Advances in metallothionein structure and functions. *J. Trace Elem. Med. Biology.* 2005, 19, 13–17.

Veglio, F; Beolchini, F. Removal of meatls by Biosorption: A review *Hydrometallurgy.* 1997, 44, 301-316.

Vijayaraghavan, K; Jegan, J; Palanivelu, K; Velan, M. Biosorption of copper, cobalt and nickel by marine green alga Ulva reticulate in a packed column. *Chemosphere.* 2005, 60,-419–426.

Vijayaraghavan K, Yun, YS. Utilization of fermentation waste (Corynebacterium glutamicum) for biosorption of Reactive Black 5 from aqueous solution. *J. Hazard. Materials.* 2007, 141, 45–52.

Volesky, B; Kuyucak, N. Biosorbents for recovery of Metals from industrial Solutions. *Biotechnol. Letters.* 1988, 10, 2, 137-142.

Volesky B.L Biosorption and Me. *Water Research.* 2007, 41, 4017–4029.

Wang, YT; Xiao, C. Factors affecting hexavalent chromium reduction in pure cultures of bacteria. *Water Research.* 1995, 29, 2467–2474.

Watanabe, A; Yano, K; Ikebukuro, K; Karube, I. Cyanide hydrolysis in a cyanide-degrading bacterium, Pseudomonas stutzeri AK61, by cyanidase. *Microbiology.* 1998, 144, 1677 – 1682.

WHO, Environmental Health Criteria 224: Arsenic and Arsenic Compounds. World Health Organization, Geneva. 2001. (http://www.inchem.org/ documents/ehc/ehc/ ehc224.htm// 1.7).

White, C; Gadd, GM. A comparison of carbon/energy and complex nitrogen sources for bacterial sulphate-reduction: potential applications to bioprecipitation of toxic metals as sulphides. *J. Indust. Microbiology.* 1996,.17, 116– 123.

Whiteley, C; Heron, P; Pletschke, BI; Rose, PD. Tshivhunge AS; van Jaarsveld, FP, et al. The enzymology of sludge solubilisation utilizing sulphate reducing systems. Properties of proteases and phosphatases. *Enzyme Microbiol. Technology.* 2002, 31(4),419–425.

Whiteley, CG; Lee, D-J. Enzyme technology and biological remediation. *Enzyme and Microbial Technology*. 2006 38, 291–316

Wilde, EW; Benemann, JR. Bioremoval of Heavy Metals by the Use of microalgae. *Biotech. Advances*. 1993, 11, 781-812.

Wiseman A. Introduction to principles. 3rd edition. In: Wiseman A, (ed). Handbook of enzyme biotechnology., Padstow, Cornwall, UK: Ellis Horwood Ltd. T.J. Press Ltd.; 1995, 3–8.

Willsky, GR; Malamy, MH. Effect of arsenate on inorganic phosphate transport in *Escherichia coli*. *J. Bacteriolology*. 1980, 144 (1), 366–374.

Wu, F; Ozaki, H; Terashima, Y; Imada, T; Ohkouchi, T. Activities of ligninolytic enzymes of the white-rot fungus *Phanerochaete chrysosporium* and its recalcitrant substances degradability. *Water Sci Technology*. 1996, 34, 69–78.

Wu, YM; Taylor, KE; Biswas, N; Bewtra, JK. A model for the protective effect of additives on the activity of horseradish peroxidase in the removal of phenol. *Enzyme Microb. Technology*. 1998, 27, 315–322.

Yamaguchi, T; Ishida, M; Suzuki, T. An immobilized cell system in polyurethane foam for the lipophilic micro-alga *Prototheca zopfii*. *Process Biochemistry*. 1999, 34, 167–171.

Yewalkar, SN; Dhum, KN; Sainis JK. Chromium(VI)-reducing Chlorella spp. isolated from disposal sites of paper-pulp and electroplating industry. *J. Appl. Phycol.*2007, 19, 459-465.

Yao, R-S; Sun M; Wang, C-L; Deng, S-S. Degradation of phenolic compounds with hydrogen peroxide catalyzed by enzyme from Serratia marcescens AB 90027. *Water Research*. 2006, 40, 3091–3098.

Zhang, X; Luo, S; Yang, Q; Zhang, H; Li, J. Accumulation of uranium at low concentration by the green alga Scenedesmus obliquus. *J. Appl. Phycology*. 1997, 9, 65–71.

In: Environmentalism: Environmental Strategies… ISBN: 978-1-60741-634-0
Editors: E. Kahraman and A. Baig, pp.101-128

Chapter 4

INTEGRATING ENVIRONMENTAL STRATEGIES INTO INDUSTRIAL PRODUCTION AND SUSTAINABLE DEVELOPMENT*

Shaoping Kuang and Hong Chen
College of Chemistry and Molecular Engineering,
Qingdao University of Science and Technology,
Qingdao 266042, PR China. E-mail: kuang1009@126.com.

ABSTRACT

In recent years, the relevant national ministries and departments and media have increased the publicity of environmentalism and its ideas. Because of environmental deterioration and its negative consequences, Environmental respect has become a need rather than just an idea. But we have verified that market forces are insufficient to regulate the impact of companies. For this reason, it is necessary to articulate a set of norms that consider the peculiarities of each industry and force companies to minimize the environmental impact of their activities. Efficient environmental protection requires global, as well as regional ecological strategies. During the last decade numerous attempts at regional and global environmental protection have been made. But industrial environmental protection is still, above all, a national matter. For China, which is now in the period of accelerating industrialization and urbanization, the per capita resource amount is below than the world average. So the circular economy will become an important practice for the country to sustain its socio-economic development. And according to the China Council for International Cooperation on Environmental Protection and Sustainable Development Task Force on Circular economy and Cleaner Production, Cleaner Production is the cornerstone of Circular economy. If China wants to vigorously build a resource-saving and environmental-friendly society, Cleaner Production should be an essential part of any comprehensive pollution management system, at the enterprise or the national level. This article puts forward some key strategies for promoting Cleaner Production in China, for instance, integrating CP into sustainability strategies, technology innovations, and industrial ecology. Lack of capacity

* The study is jointly supported by the Qingdao Social Scientific Foundation of China (No. QDSKL070216) and The Key Project of QUST Social Scientific Foundation, China (No. 07SKA19).

to track and evaluate the effects of CP programmes, and to disseminate this information, still seems to be one of key challenges for CP initiatives. So development of a Cleaner Production indicator system and recommendation of technical criteria is essential. How we design, how we meet people's needs, how we use products, and leading to excessive consumption are some of the challenges we face to reach a sustainable and sustaining future. The efficient solution will require social, economic and technological innovations. In this article, Economic Policies for Sustained Development is described, and advocated as an important first step in a strategy moving towards economic-cum-environmental sustainability. Drastic changes and bold initiatives are suggested and required to establish true sustainable development for China.

Keywords: environmental strategies; ecological strategies; Cleaner Production; circular economy; industrial ecology; sustainable development.

INTRODUCTION

China's eleventh five-year plan of national environmental protection has put forward energy consumption of per unit GDP would decrease about 20 % and the confinement indices of main pollutants emission reduce 10 % during eleventh five-year. Two confinement indices of eleventh five-year pollution emission reduction are: in 2010, chemical oxygen demand (COD) and SO_2 emission quantity decrease 10 % compared with 2005 respectively, that means the national COD which is 1414.2 million tons in 2005 will reduce to 1272.8 million tons, and SO_2 emission quantity which is 2549.4 million tons in 2005 will reduce to 2294.4 million tons.

In 2007, due to the correct leadership of central committee and state council, each region and sectors have thoroughly implemented the Scientific Concept of Development, regarded pollution emission reduction as an important task, and adopted comprehensive treatment and pollution control. So the push forward dynamics further increase, policies and measures further carry out, and pollution emission reduction gains breakthrough progress. COD and SO_2 emission quantity realize double decrease, which show that environmental protection work obtains significant developments. The national COD is 1381.8 million tons, which decreases 3.2 % compared with last year, and SO_2 emission quantity is 2468.1 million tons, which decreases 4.7 % compared with last year. The national COD and SO_2 emission quantity decrease 2.3 % and 3.2 % respectively, compared with 2005 (2007 Report on the State Environment in China; Liu and Chen 2007; Liang and Sun 2007).

Emission of waste water and main pollutants in recent years is listed in table 1. In 2007, the national total quantity of waste water was 556.7 billion tons, which was increased 3.7 % compared with 2006. Emission of COD was 1381.8 million tons, decreased 3.2 % compared with 2006. Emission of NH_3-N was 132.3 million tons, decreased 6.4 % compared with 2006.

In 2007, SO_2 emission quantity was 2468.1 million tons, smoke dust emission was 986.3 million tons, and industrial dust emission was 699.0 million tons, which were decreased 4.7 %, 9.4%, and 13.5 % respectively (table 2).

Table 1. Emission of waste water and main pollutants in recent years

Items / year	quantity of waste water (billion tons)			emission of COD (million tons)			emission of NH_3-N (million tons)		
	life	industry	total	life	industry	total	life	industry	total
2005	281.4	243.1	524.5	859.4	554.8	1414.2	97.3	52.5	149.8
2006	296.6	240.2	536.8	885.9	542.3	1428.2	98.8	42.5	141.3
2007	310.2	246.5	556.7			1381.8	98.3	34.0	132.3

Table 2. Main pollutants emission in the waste gas in recent years

Items / year	SO_2 emission quantity (million tons)			smoke dust emission (million tons)			industrial dust emission (million tons)
	life	industry	total	life	industry	total	
2005	380.9	2168.4	2549.3	233.6	948.9	1182.5	911.2
2006	351.2	2237.6	2588.8	224.3	864.5	1088.8	808.4
2007	-	-	2468.1	215.5	770.8	986.3	699.0

Table 3. Emission sources of direct discharge in sea

emission sources	wastewater (billion tons)	COD (million tons)	petroleum (tons)	NH_3-N (tons)	P (tons)	Hg (tons)	Cr^{6+} (tons)	Pb (tons)	Cd (tons)
industry	10.97	4.09	467	2375	196.3	0	2.80	0.69	0.45
life	6.90	9.00	970	10179	1366.3	-	-	-	-
others	23.72	28.40	1405	38006	3250.2	0.23	24.33	13.79	1.79
total	41.59	41.49	2842	50560	4812.8	0.23	27.13	14.50	2.24

National environmental quality generally shows better tendency in 2007 from table 1 and table 2. But still there are many problems needing urgently to be studied and solved. The pollution situation of surface water still severe: seven water systems are moderate pollution in general, offshore area is slight pollution in general, and the problem of lakes eutrophication is increasingly outstanding. Pollution in the nearshore waters is a major environmental issue in China (table 3).

We can see total wastewater discharge of 607 emission sources of industry, life and others, which are more than 100 tons daily wastewater discharge, was 41.59 billion tons. Total emission of COD, petroleum, NH_3-N, P, Hg, Cr^{6+}, Pb, Cd were 41.49 million tons, 2842 tons, 50560 tons, 4812.8 tons, 0.23 tons, 27.13 tons,14.50 tons, 2.24 tons, respectively. How to dominate the water environment and ensure economic growth simultaneously is an urgent issue to be tackled at present.

Consumption rate of plant resources is too high and fast because of rapid development of economic society and population growth. Endangered species rise quickly and about 4000~5000 kinds of higher plants are deteriorated or in threatened status. The quantity of threatened plant species is about 15 % ~20 % of the whole higher plants species in China, which is higher than the global average level.

Environmental protection situation in rural areas is still severe. The coexistence of point sources pollution and plane sources pollution, the superposition of domestic and industrial pollution, the interweave of new pollution and old pollution are the most important factors which should be absolutely bring about serious concern and reaction of the government and related institutions of China. Industrial and urban pollution transferring to rural pollution is a crisis of the safety of drinking water and agriculture products. The dual threatening including environmental pollution and ecological destruction restricts the development of economy, agriculture and ecosystem in rural areas (Wei, liu et al 2007; Huang, Gao et al 2008).

So we should continuously improve our environmental programs and find solutions to our environmental challenges. We've focused our environmental sustainability efforts on water, energy, packaging and solid waste — areas where we can make the biggest impacts. In addition, we have many programs in the area of sustainable agriculture and we partner with our growers. We encourage conservation, recycling and energy use programs that promote clean air and water and reduce landfill waste. To form complete, harmony and sustainable development principle and to realize sustainable development, we must change the old mode used by many developed countries that is the one " pollute first, manage later", which is the industrialization based on depleting resources and destroying environment, and must deal with the relations between economic development, population, resources and environment correctly.

Environment and Economy Effect Analysis of Circular Economy to Environmental Kuznets Curves

The relationship of environment and economy is the focus of attention in theory circle. There are two important relatively separated academic theories. First, an assumed theory, which reveals the trajectory of development and change of economy and ecological environment, represents a phenomenon and embodies a transformation of contradiction. That is to say ecological and environmental situation and economic development level follow Environmental Kuznets Curves which presents the shape of inverted U. Another academic theory is unification of theory and practice, that is to say economy and environment can realize the positive interaction new idea in circular economy which reveals essentially the internal relations of environment and economy and realizes one leap in environmental economics thought. Environmental Kuznets Curves has proved by historical track of developed countries and the circular economy model is also now exploring and practicing by many countries. Although these two researches have different forms and emphasis points, they don't divorce from each other or conflict with each other. This new economic model – circular economy not only agrees with Environmental Kuznets Curves, but also represents the interactive relationship between economy and environment. Therefore circular economy is the optimal development model which can realize the interaction of economy and environment.

The circular economy theory can be considered as an improvement to the contents of the Environmental Kuznets Curves and the practices of circular economy will make the Environmental Kuznets Curves more perfect (Zhang 2005).

An Analysis of Environmental Kuznets Curves

Environmental Kuznets Curves is established in a coordinate, in which ordinate represents the ecological degradation degree of environmental quality, abscissa represents the increase situation of wealth per capita. The left curve of point A in the coordinate shows the early industrialization development stage which had lower economic level and less effect degree on environment (figure 1). The curve rises slowly until the accelerating period of industrialization process or economic growth. The consumption of natural resource and reduction quantity of scrap and exhaust gradually exceed the environmental self-purification capacity. Ecological environmental quality deteriorates quickly with the rapid increase situation of wealth per capita. Non-sustainable of environment and economy increases constantly. The right and nearby curve of point A in the coordinate shows transformation period of post-industrial society or economic development in which wealth per capita has the trend of continuous increase. Ecological environmental quality gradually improves after a period of stable state and plateau phase. Sustainable of environment and economy shows a tendency of strengthening.

Environmental Kuznets Curves reveals the interaction of environment and economic development. The left curve of point A in the coordinate mainly reflects a malignant interaction, which means the mankind tries to realize the economic benefits at the expense of ecological environmental quality. The right curve of point A in the coordinate mainly reflects a relation of beneficent interaction in which economy and environment gradually represent the harmonious development (Fan and Hu 2002; Hong and Luan 1999; Grossman and Kreuger 1995; Davia, Rebecca et al 2000; James and Arik 2001).

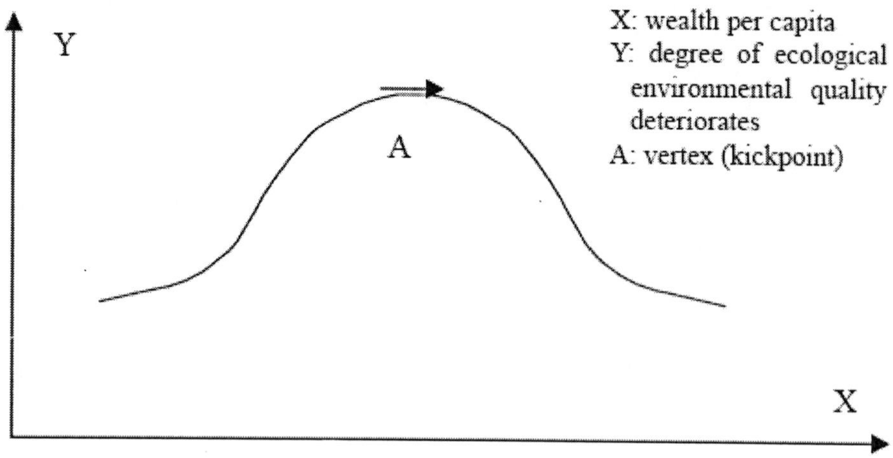

Figure 1. Environmental Kuznets Curves schematic diagram.

The changing process of Environmental Kuznets Curves incarnates a dynamic interactive influence process in the development of environment and economy. People influence environment through various activities especially economic activities, while environment is the basis which human existence and development rely on (Torras and Boyce 1998). Therefore, it is important for the improvement of Environmental Kuznets Curves to harmonize the relationship of environment and human economic activities. And establishing a benign dynamic interaction is the best way to realize this.

Benign Effect Analysis of Circular Economy to Environmental Kuznets Curves

Circular economy is the optimal development model which can realize the interaction of economy and environment at present. The dynamic interaction of Environmental Kuznets Curves embodies the transform of contradiction between human economic activities and ecological environment. Traditional economy is a linear economy growing way which is constituted with "resource-products-pollution and discharge" of substance flowing in a single direction. Human realize quantity growth of economy through extensive operation and disposable using to resources in this economic operation. Various phenomena such as high consumption, high output and high waste directly cause the deliberate destroy of the ecological environment. The contradiction between the economic activities of growth-type mechanism with the limitlessness of resources requirement and the ecological environment of stable type mechanism with the limitation of resources supply is the basic conflict in traditional economy.

This contradiction has two aspects: one is economic growth, which is an objective requirement of the development of human society productivity; another is the relatively stable ecological environment, which is an objective trend of the development and succession of ecological environment. The first half curve of Environmental Kuznets Curves embodies the plunder economic growth and the ecological environment resources assumption to meet the need of economic development in human society. Thus, the contradiction between human ecological activities and ecological environment activates and sharpens continuously (Romon 2000). After several decades of seeking and practices around this more and more outstanding contradiction, human find a brand new development model-circular economy model until the late 20th century (Li and Bu 2003).

Circular economy simulates the operation mode and law of natural ecosystem, and emphasizes transformation from only reuse to reduce, reuse and recycle. It has changed the linear traditional economy "resources—products—waste" to the process of "resources–products – recycling resources". Therefore it is a way of sustainable production and consumption pattern at the inevitable request of new industrialization. In the narrow sense, circular economy refers to reduce of wastes and resource recovery. It can fall into the category of "rubbish economy" or "waste economy". In the broad sense, the circular economy covers all the socio-economic activities. All materials and energy are used rationally and continuously in sustained economy cycles, and the harmful effects of emissions and pollutants can be reduced to a minimal level. The way of circular economy can alleviate environmental pressure of economic growth to realize persistent support of environment and resources to economic construction. Circular economy model is an inevitable choice because of intensification of the contradiction between human economic activity and ecological environment (Zhou 2006). Circular economy and market economy have identity. As a brand new model to realize the recycling of substances and energy in economic development process, circular economy utilizes natural resources and environmental capability with an environmental friendly way to protect environment and economic development and realizes ecological trend of economic activities with less cost and higher efficiency. Market economy is an economic operation mode which takes market as basic means of resources optimization development and is started and regulated by market mechanism. At the condition of market economy, all economic activities are directly or indirectly in market relationship, and all labor products and production factors are collocated by market. The purpose of producer's production is realizing the maximum of its profit. What to produce, how many products are produced and how to produce are regulated by the power of supply and demand and price mechanism. Market mechanism is a basic operation mechanism which can promote the flow of production factors and optimal allocation of resources (Xie 2004). Both circular economy and market economy pursue efficient utilization and optimal allocation of resources. Fundamentally, they are according with economic development and they have identity even though there are different reflections in economic development idea and model. The identity of circular economy and market economy indicates circular economy model is an excellent selection to alleviate the contradiction between human economic activity and ecological environment.

If we can consciously adjust our economic behavior and economic way when we recognized the essential relationship of environment and economy, there must have good effect on Environmental Kuznets Curves. Point Q in the left curve of Environmental Kuznets Curves not only indicates a process point, but also predicates a new start point. If unceasingly using the traditional economic pattern, environment will worsen continuously. But if we can implement circular economy as soon as possible, and extract the experiences and lessons of developed countries, it is completely possible establishing a benign and unique Environmental Kuznets Curves (Zhao, Li et al 2005). The significance of implementing circular economy lie in establishing the relationship of mutual promotion of economy and environment, and leaping the peak of Environmental Kuznets Curves with relatively smaller economic scale and lower pollution level which is embodied in Environmental Kuznets Curves that the peak left shifts in horizontal direction and down sifts in longitudinal direction (figure 2).

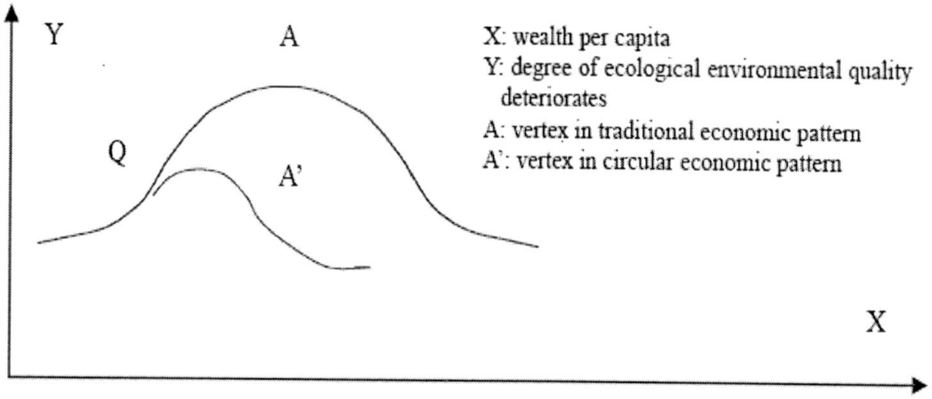

Figure 2. The comparative schematic diagram of Environmental Kuznets Curves under circular economic pattern.

Effect Analysis of Circular Economy Practices to Environmental Kuznets Curves

The circular economy theories enrich the connotation of Environmental Kuznets Curves, and influence its trajectory from microscopic to macroscopic practice. The process of leaping the peak of Environmental Kuznets Curves reflects tortuous experiences in the exploration and stride of circular economy mode which changes the traditional economy.

The background resulting in circular economy in Germany and Japan is various (List 1999; Hilton, Levinson et al 1998; Selden and Song 1994). In Germany, direct motivation is the realistic restriction of land space in adopting traditional landfill way to deal with waste. Additionally, resources shortage impels the country stepping the road of cycle which aims at the waste of social consumption. While in Japan, resources shortage especially oil crisis in 1970s, has deeply promoted energy efficiency in national industrial production. Although backgrounds and causes are multiplicity, the essential factor is that these countries realize the irreconcilable disadvantages in the process of industrial modernization with environmental damage and predatory consumption of resources and energies. In 1770s and 1980s, developing countries start putting forward circular economy and carrying out the practice in microcosmic field and area coverage. At the same period, these countries start leaping or have leaped the peak of Environmental Kuznets Curves. Developed countries leave an underway trace of circular economy in the trajectory of Environmental Kuznets Curves, and circular economy exploration and practice also promote the process of leaping the peak of Environmental Kuznets Curves.

The degree of exploration and effort directly influence the position of point A in Environmental Kuznets Curves. American, Germany and Japan arrive at the peaks with 11000 US dollars per capita GDP, 10000 US dollars per capita GDP and 8000 US dollars per capita GDP, respectively. American is comparatively late which is because of the more effort in Germany and Japan than that in American in the implementing of circular economy. Also there are some post-developmental countries and areas which start improving economic mode from relatively lower economic levels. Thereby they realize the practice with even lower level in leaping the peak.

The significance of circular economy to Environmental Kuznets Curves in China is bigger than that in developed countries. Developed countries have completed the process of industrialization via high resource consumption, while China being in the process of industrialization should require design to the development process. Because of the shortage of nature resources and limited environmental capacity especially in the background of economic globalization, rapid development of information industry and technological innovation, China has a larger space and rich content in promoting the circular economy (Yang, Ye et al., 2003; Qu 2006; Chen and Cai, 2007) . Circular economy, which can improve resource utilization efficiency, reduce pollution cost in the production process, and decrease ecological destruction in the consumption process, will certainly greatly improve Environmental Kuznets Curves in China's development.

Environmental Management Strategies in Industries

The Environmental Management System

Continuing and expanding industrial development is essential in most developing countries, and it is timely to establish meaningful controls on industrial pollution emissions before the problem gets out of hand. Up to now, in most developing countries, attention to industrial waste management has not been taken very seriously. This is probably because of the interest of the decision makers in promoting industrial growth, together with the lack in many developing countries of meaningful enforcement systems. The absence in virtually of effective monitoring systems for observing and documenting industrial performance in managing wastes is noticeable, despite the fact that in the industrialized countries it is axiomatic that "without effective monitoring it is not possible to have effective control".

The solution to industrial pollution problems could be readily accomplished, if there was the will to do so, because the solution is economic and affordable. Experience worldwide has shown that it is essential to establish the "Industrial Waste Permit System" by which every industry which produces significant wastes is required to install and operate its own treatment system before discharging waste effluents to the environment (to a drainage ditch, or to a municipal sewer, or to a joint industrial waste treatment facility). Each industry should pay fees for its share in the cost of administering the permit system and this should include the cost of monitoring. Experience in the USA and other industrialized countries has shown that this does not impose costs on industries which harm their competitiveness provided that the requirements for treatment are appropriate, that is, based on appropriate standards tailored to most actual protection needs. So establishing an appropriate industrial waste permit system and the monitoring to go with it are essential because of the common and effective use of this method of control in the industrialized countries and the availability and practicability of the technology.

Keeping in mind the interest of our future generations, sustaining harmony with the environment in the process of conducting business activities is one of the Management's top priorities (A World Bank Policy Research Report). Activities, which have significant environmental aspects and impact were systematically identified and evaluated with control measures in placed to minimize the environmental impact. Various environmental management programs were also implemented to reduce environment impact caused by business operations. The Environmental Protection Agency (EPA) was often under siege, financially as well as in regulatory issues. But this phase was not exclusively oriented towards minimizing the regulatory and financial burden on industry and business for pollution and energy use. So the shift in policy orientation was based in the assumption that the debate was no longer chiefly over the appropriateness of having environmental safeguards. What is most needed is the right opportunity and incentives for business and industry to marshal their creativity and technological know-how to meet the needs of a less-polluting and more energy efficient society" (Mazmanian and Kraft 1999).

Objectives and targets are controlled through the Environmental Management Programmes. Project leaders are appointed for establishing the Environmental Management Programmes to meet the objectives and targets set for the activities under their controls. The status and results of the programmes are reviewed monthly during EMS (Environmental Management system) monthly review meeting and countermeasures taken whenever necessary.

All activities, products and services carried out in the company which has significant aspects and impact to the environment has been systematically identified and evaluated in accordance to the established procedure. For example, Green Factory Activities: The Energy and Water Conservation Project and Zero Emission Program are two such environmental management programs being executed to minimize use of resources and reduction of industrial waste; Eco–Product: being environmental friendly has always been the Management top priority when it comes to product development, starting from the sourcing of materials to the making of the completed product. We have taken careful steps to ensure that the materials used in our products are environmental friendly and our finished products come with low power consumption function features. The company should make stage goals to evaluate the implementation of environmental work, for example, short-term objective, mid-term objective, and long-term objective. Some characteristic disagreements and contradictions are in a similar mode environmental policy and politics. First of all no consensus can be reached on the balance between long-term and short-term environmental issues. Just as it is a dangerous orientation of science policy to concentrate on short-term and merely economic output, it is inadequate to direct environmental policies on mitigation of the worst pollution and degradation without taking in focus the long-term processes-which result from apparently non-hazardous normal practices as well as from technological innovations (Reller 2000).

Table 4 shows the stage goals of a golden company. This company which relies on the advanced production technology and management technology efficiently implements resources circulation utilization. The strategic development of the golden company are following: Energy consumption and water consumption of industrial output value, comprehensive recovery rate and so on reach advanced level in domestic, enterprise comprehensive competitiveness and ecological level are in a leading position in gold industry, and the company steps into the track of sustainable development.

Table 4. The stage goals of a golden company

category	number	index	unit	present situation	2010	2015	2020
economic and technical indexes	1	gross industrial output value	%				
	2	production scale	ton/day	1400	1500	1700	2000
	3	complex energy consumption of total industrial output value per million yuan	tons of standard coal	0.84	0.8	0.6	0.5
	4	fresh water consumption of total industrial output value per million yuan	m^3	26.7	25	22	20
	5	electricity consumption per unit ore	Kwh/t	59	50	42	35
	6	mining loss rate	%	7.41	7.0	5.5	4
recycling indexes	7	ore recovery	%	92.9	93.5	96	98
	8	dressing recovery	%	94.8	95.5	98	100
	9	total recovery of beneficiation and metallurgy	%	92.5	94	96	98
	10	utilization ratio of wastewater cycle	%	70	90	100	100
	11	wastewater treatment rate	%	80	100	100	100
	12	utilization ratio of tailings disposal	%	33	90	100	100
	13	utilization ratio of smelting tailings disposal	%	90	95	1010	100
ecological management indexes	14	green cover percentage in mining area	%	8	10	15	20
	15	percentage of enterprises implementing cleaner production	%	/	33	100	100
	16	pass rate of ISO14000	%	/	33	1010	100
	17	cognition rate of circular economy	%	60	100	100	100
	18	environmental investment percentage	%	2.43	2.8	3.5	5

The key characteristics of the company's operation and activities that have significant impact on the environment are identified and the frequency determined according to the legal requirements. The significance of our environmental aspects and associated impacts is evaluated under normal, abnormal and emergency conditions on the basis of a series of weighted factors to assess the environmental risks which include environmental factors and business factors, including regulatory requirements. Examples of the environmental performance being monitored and measured are: air emission, trade effluent discharge, noise level and industrial hygiene. The performance is measured against the environmental target and/or regulatory specification. Test results are checked and evaluated for compliance and reported during regular review meetings. Rectification actions are taken whenever test results exceed the set parameters and retest will be conducted to ensure results met the control limits. Whenever there is an introduction of a new process or installation of new equipment or any change in activity, service, product or environmental regulations, the results of the aspect evaluation shall be reviewed to ensure the ratings are still appropriate and up-to-date.

Ecological Management System

Ecological management is an embodiment of mechanism innovation and management innovation in the industry which is taking a new road to industrialization. Construction and management must embody the principle of ecological management. In the process of construction and operation, company should implement green management and obligation service system based on the international ecological environment management system ISO14000. Selecting project investment according to industrial ecosystem, establishing ecological environment management system, guiding cleaner production, energy-saving, waste exchange, and comprehensive utilization of resources, ecological environment management system authentication, these are what the company should do to improve green infrastructure of material, energy, and information and provide material and technical support for ecological-system establishing and healthy operation.

Under the guiding of the framework system of ecological management (figure 3), in the park area, the strategies and policies of ecological environment management mainly include: ecological environment management system ISO14000 in group, green infrastructure construction, APELL plan, etc. In the enterprise, the strategies and policies of ecological environment management mainly include: ecological environment management system ISO14000 in enterprise, cleaner production, waste minimization, resources recycling, etc. In product, the company should actively implement the policies of environmental park and life-cycle management to products. Establishing product-oriented environment strategies and enterprise environmental management system is the basic prerequisites of implementing green design.

For the management of industrial wastes, effective monitoring and enforcement, including use of appropriate effluent standards, will be required. New industries (and some existing ones) should be sited to utilize areas where waste pollution management will be simplified. This will involve the use of industrial estates.

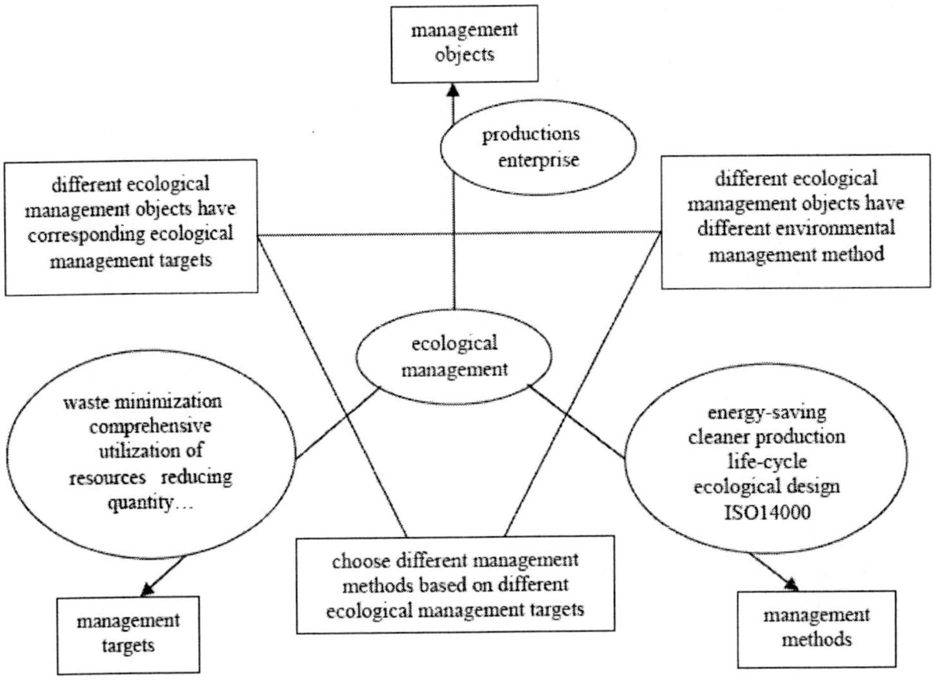

Figure 3. the framework system of ecological management.

The controlling and management of industrial solid waste should establish enterprise environmental management system and carry out cleaner production according to the reduction, resource and harmless thinking of solid waste. Classified collection of solid waste is important for resource project construction of solid waste, and the construction of ecological industry chain can increase the utilization efficiency of resource of solid waste. According to the national law of solid wastes environmental pollution control, a systematic study has been made on the treatment and disposal technology for industrial hazardous wastes. Integrating innocuous treatment plants and security landfill are effective methods to implement the harmless treatment of industrial solid waste (figure 4). Meanwhile, environmental protection department should establish the trade management system, marketization mechanism and the price system of waste treatment which include charge system of waste treatment and management system of operation to promote the marketization of waste treatment.

Large quantity, complex chemical ingredients and noxiousness, these are the characteristics of industrial solid waste. Their control approaches have three steps: first is reduction, which means improving production technology to reduce the waste residue and the loss amount of raw materials; the second is resource, which means recovering useful materials in the waste residue and manufacturing by-product using waste residue; the third step is harmless, which makes the final treatment to the waste residue after recycling combining with urban construction, health and environmental protection departments, and the final emission must reach national emission standards.

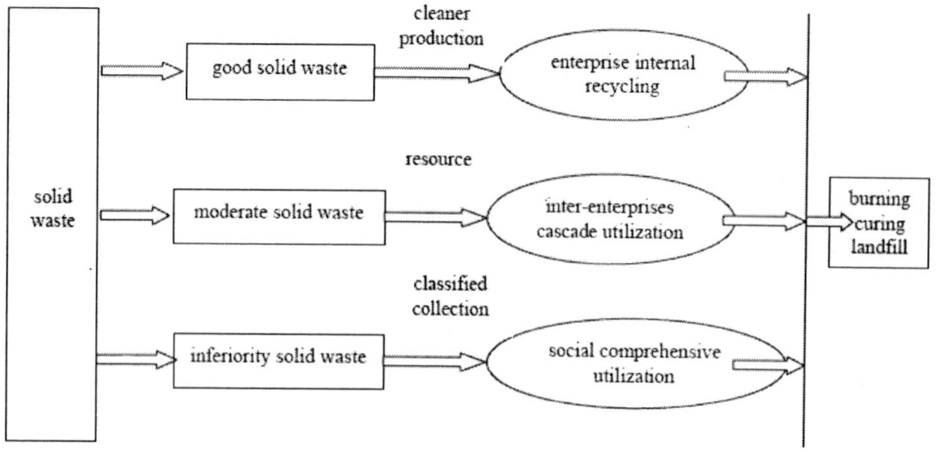

Figure 4. Comprehensive management of solid waste.

(1) Reduction

The reduction of industrial solid waste is mainly realized by implementing cleaner production in industries. The main measures are as follows:

- select suitable production raw material and implement cleaner production technology combining with technical transformation to eliminate or reduce waste production from source.
- select high quality raw materials and mechanical equipments to improve product quality and service life, reduce the waste quantity in the production process.
- realize comprehensive utilization of waste through enterprise internal recycling and inter-enterprises cascade utilization, recycle leftover materials and integrate raw materials by simple treatment which can reuse in the production process.

(2) Resource

The planner of enterprise should construct series of ecological industry chain through establishing by-products exchange system in enterprise or between enterprises to reach gradient utilization of solid waste and implement recycling and reusing of by-products or materials. Therefore, the enterprise can recover resources furthest and reduce the final emission quantity of waste which can give full play to utilization value of resources.

If enterprises want to establish resources network of industrial solid waste, they must analyze the characters of solid waste from the industries, the relationship between upstream and downstream and the feasibility of technology and economy. Then they can establish a comparatively perfect metabolic system of industries waste through waste metabolic network and series of ecological industry chain in enterprise or between enterprises and the symbiotic cooperative relationship between by-productions and materials in enterprises.

(3) Harmless

The enterprises should make specialization treatment to the intractable harmful waste combining special environmental protection company and related government facilities.

Cleaner Production

From 1993, Chinese government started to advocate Cleaner Production, and now it is pushing for the Circular economy. All of these are the unremitting efforts that our country is making to realize a sustainable development. Some may simply regard Circular economy as waste recycling, yet the fundamental goal is to systematically prevent and reduce wastes in the industrial process. And according to the China Council for International Cooperation on Environmental Protection and Sustainable Development Task Force on Circular economy and Cleaner Production, Cleaner Production is the cornerstone of Circular economy. If China wants to vigorously develop the Circular economy and build a resource-saving and environmental-friendly society, Cleaner Production should be an essential part of any comprehensive pollution management system, at the enterprise or the national level.

Cleaner Production started as an integrative and preventive environmental management strategy for avoiding, or at least minimizing, environmental impacts from industrial products and services, as well as from the production, distribution and service processes they require. The field has matured over the last 25 years. During the last decade, in particular, there has been remarkable progress putting Cleaner Production on the agenda of industry, government and communities in both industrialized and industrializing countries. Cleaner Production programmes have generally been successful in demonstrating potential environmental, financial and other benefits of an integrated, preventive environmental management strategy for industrial operation (Hammer 2001).

In recent years, the relevant national ministries and departments and media have increased the publicity of Cleaner Production and its ideas. After the central government proposed the thought of scientific view of development which stressed the principle of people first and coordinated and sustainable growth, the public has been better aware of the significance of the Cleaner Production.

Cleaner Production is a 'win-win' strategy. It protects the environment, the consumer and the worker while improving industrial efficiency, profitability, and competitiveness. The less businesses waste, the more profitable they can become (UNEP Environment Programme).The initial means to prevent industrial pollution was to "end-of-pipe" control, which was a measure that only input but not yielded economic output. Experiences and lessons from practices around the world have proven that end-of-pipe treatments are not a cost-effective way of reducing industrial pollution. Instead, Cleaner Production, first developed in the 1970s, has obtained both environmental and economical benefits in different countries. Although Cleaner Production is based on a preventive mindset, current Cleaner Production efforts are often still made in the final phases of industrial development rather than being designed in right from the start. In other words, Cleaner Production has moved industrial environmental management from the end of the production pipe to the end of the innovation pipe, but has not yet been integrated into the innovation cycle as required for the transition towards a priori clean products and processes (Van, Willems et al 1997).

Technology Innovations: A Trump of Cleaner Production

Most importantly, Cleaner Production requires a change of attitudes and technology improvements. Advances in science and technology open up new opportunities and options for achieving CP. Approaches to CP can be grouped into three categories (John and David 2001):

- Level 1: Waste reduction at source good housekeeping; process modification; product modification; change of materials.
- Level 2: Waste recycling internal recycling; external recycling (among different organizations industrial ecology).
- Level 3: Use of renewable resources biomass as a renewable feedstock for energy, fuels and chemicals (www.carbohydrateeconomy.org); other sources of renewable energy: solar, wind, tidal, small-scale hydroelectric.

These approaches involve a combination of technology (processes and tools) and techniques (ways that technology is used). Technology and techniques are mutually supporting in delivering CP, as seen in case studies. Performance assessment is the guide to determining the optimum Chemicals and energy are critical sectors for CP technology in view of their importance to so many other industry sectors. Biotechnology also holds great promise for achieving sustainable industrial development. Most, if not all, technologies can be relevant to CP. Energy is a fundamental requirement for human activities, from cooking to industrial manufacturing to transportation and travel. The predominant energy source at present is fossil carbon, specifically oil, natural gas and coal. The industry has invested for decades, and continues to invest, in technology to divert/convert more waste into product, as well as to replace polluting chemicals and develop new more environment-friendly products (www.shell.com/royal-en/).

The chemical industry is a key to advancing CP, as chemical products and processes are used as inputs in so many other sectors. Chemical technology has increased in efficiency, with decreased waste as a secondary benefit. This change has largely been driven by competition and regulation. The reduction in waste per tonne of product has been accomplished by approaches ranging from improved housekeeping to improved processes and process control, development of new less polluting products, and more efficient utilization of by-products. Technology innovation has occurred in the areas of catalysts, sensors and reactor design, as well as materials and separations technologies. Recently the concept of Green Chemistry or "environmentally benign chemistry" has shifted the focus to reducing the hazard of chemical products and processes, and to controlling emissions and exposure. Industrial biotechnology can be viewed as a subset of Green Chemistry involving bioprocesses that are more eco-efficient, and biologically derived chemicals that are more biodegradable than their conventional chemical counterparts. Biotechnology has begun to play an important role in the synthesis of fine chemicals to the extent that biocatalysis is becoming part of the general tool kit used by synthetic organic chemists, often providing environmental benefits. For example, biochemie, a subsidiary of Novartis, has reported that for the production of one tonne of cephalosporin the conventional chemical process resulted in 31 tonnes of wastes requiring incineration, while its new enzyme biocatalysis process resulted in only 0.3 tonnes of such wastes (OECD). On the horizon are bioprocesses to produce a number of bulk and commodity chemicals as well as plastics utilizing biomass as a renewable feedstock. Bioprocesses naturally lend themselves to processing of biomass, a major renewable resource, into fuels (e.g. methane, ethanol) and chemicals (e.g. acetone, butanediol).

Technology innovation can be a very powerful tool for achieving CP if it is informed by effective technology assessment and supported by effective policies at the company level. A wide range of technologies and related research disciplines can contribute to CP. However, advances in science and technology are not sufficient as a driver for achieving CP. Market forces and government policies must also work in concert with technology to advance industry's adoption of CP. Thus, a wide range of different communities representing quite different perspectives must come together to help achieve CP. One integrating principle for CP, which focuses on sustainability, is use of CP technologies to establish a sustainable linkage between the carbon cycle in industry and the carbon cycle in the environment. Technologies based on life sciences and biotechnology, in particular, will play an increasingly prominent role in moving global production systems toward this ultimate goal of sustainability.

Industrial Ecology: A New Cleaner Production Strategy

Communicating the fact that end-of-pipe technologies are more expensive is the educational challenge for Cleaner Production advocates. At a more profound level, industry and governments need to achieve an understanding of systems thinking: the ecological system is the all-encompassing one, with cascading subsystems and sub-subsystems (social system, economic system, industrial system and so on). Systems' interdependence cannot be broken. So Cleaner Production is not the domain of a single government agency, or of private sector manufacturing only; Cleaner Production and eco-efficiency remain targeted towards manufacturing processes and business strategies within industrial companies. However, Cleaner Production could be applied at the level of a cluster of various companies, an industrial zone, or even a whole region (that is, it could be applied at the level of a system). This emerging approach has become known as "industrial ecology." The ultimate goal of industrial ecology is to determine how the industrial system could be restructured to make it compatible with the way natural ecosystems function.

The industrial system can be seen as a certain kind of ecosystem. Like natural ecosystems, it can be described as a distribution of materials, energy and information flows. Furthermore, the entire industrial system relies on resources and services provided by the biosphere, from which it can not be dissociated (Frosch and Nicholas 1989; Erkman and Suren 1997) .

Thus the concept of "eco-industrial parks" (EIPs) originated in the early 1990s. EIPs are areas where companies cooperate to make the most of resource use, namely through mutual recovery of the waste they generate (waste produced by one enterprise being used as raw material by another).

The story of Kalundborg, Denmark, really begins in 1961, with a project to use surface water from Lake Tiss for a new oil refinery to preserve limited ground water supplies. The city took responsibility for building the pipeline, while the refinery financed it. A number of other collaborative projects were introduced and the number of partners gradually increased. By the end of the 1980s, the partners realized they had effectively "self-organized" into what is probably the best-known example of a working industrial ecosystem or, to use their term, industrial symbiosis.

The industrial symbiosis in Kalundborg is an example of how strategic material-based planning can earn a handsome payback. There is no doubt that the Kalundborg model has fruitfully inspired recent thinking on environmental management of industrial estates and eco-industrial networks. However, there is also growing recognition that we need to look beyond Kalundborg. This is especially true with respect to implementation of industrial ecology in developing countries, where the industrial pattern is very unlike that of Kalundborg (Gertler, Nicholas et al 1997)

Key Strategies for Promoting Cleaner Production in China

With ongoing efforts to ensure compliance and to enhance the capacity of enterprises, communities, and local officials, valid policies could become effective tools to promote Cleaner Production in China. We wish the following suggestion could be available for the relevant national ministries and departments.

(1) Incorporate Cleaner Production into Government Policies and Strategies

Continue to select and provide examples of successful integration of Cleaner Production in national/local policy frameworks. In addition, intensify work in key areas like industrial estates where there is the potential for multiplier effects.

Policy instruments that promote Cleaner Production (CP) are not fundamentally different from other policy instruments. However, they must be wisely conceived in order to favor prevention rather than end-of-pipe approaches.

During the 1990s, Cleaner Production was introduced in China, in the beginning especially via development aid projects. From 1992 to 1997 the focus was strongly on the introduction of Cleaner Production methodology, the training of personnel and the implementation of demonstration projects at the enterprise level. After 1997, the emphasis shifted to cleaner production policy-making, culminating in the Cleaner Production Promotion Law in 2002. This paper analyses the making of this unique law and provides a first assessment of the strengths, weaknesses and particularities of the first law on Cleaner Production.

Our country should make elaborate chemicals policies - which substances are restricted in acceptable emissions and which substances are urged permits regulating emissions to air and water, and waste management. It is clear to what extent licensing authorities will be able to include Cleaner Production approaches in actual permit issuing. The government should do well for this.

(2) Promote Cleaner Production Financing

CP often results in sufficient cost savings. The payback period for investment in CP technology innovation may be a few years or less. A key element for bringing in financing, whether from the company's own resources, the government or a financial institution, is building awareness of CP's economic benefits and the costs of continued pollution of the environment in the organization providing the financing. In this regard, tools such as LCA and Environmental Accounting can be useful. The use of Environmental Management Systems can also provide comfort to investors that the objectives of CP technology development will be met.

More work needs to be done to develop capacity for integrating preventive strategies in accounting and due diligence practices among related stakeholders such as financial institutions, enterprises, business schools and the media. Revolving funds should be encouraged, and governments should formulate rules and incentives to stimulate investment in Cleaner Production implementation.

Much more attention needs to be paid to issues of financing when examining technical option, at the enterprise or the sector level. This may require training both industry staff and financiers in the preparation and analysis of project proposals.

(3) Development of a Cleaner Production Indicator System and Recommendation of Technical Criteria

Develop a standard set of measurements for CP. Rhetoric from governments, industry, and others about their CP actions is high. But, we lack agreement on affordable ways to measure CP progress. Without measurement standards, people cannot be held accountable.

Cleaner production programmes have been instituted in a wide array of firms and institutions. There are many case reports on their successes. Yet these programmes lack common metrics for measuring performance. They are seldom assessed against their full costs, and there is little possibility to compare one project against another to determine the effectiveness of differing strategies. Few government programmes publish annual trend reports on their environmental impacts, and even fewer have been evaluated for cost-effectiveness. The highly contextual character of Cleaner Production programmes makes it difficult to develop common metrics or units of analysis. Without measuring performance against financial or environmental objectives, the specific impacts of cleaner technologies and practices cannot be effectively assessed.

Interest in corporate environmental reporting and sustainability indicators is growing rapidly(Bennett, Martin et al 1999). Professional bodies, including Dow Jones, the American Institute of Chemical Engineers (AICE), the Social Venture Network and the World Business Council for Sustainable Development (WBCSD), have developed environmental indicator systems for corporations. The Coalition for Environmentally Responsible Economics (CERES) is developing an ambitious Global Reporting Initiative to track corporate environmental performance. These efforts suggest how useful and feasible indicators of Cleaner Production might be; China should inspire Cleaner Production promoters to work towards developing such indicators.

(4) Formulation of Supporting Incentive Policies to Implement Cleaner Production

Government has a responsibility to provide conditions for economic development and environmental betterment. In this context, government should play a key role in the promotion of CP. Government policies that support incentive policies will be beneficial for the implementation of Cleaner Production.

Specific reforms in the energy and other public utilities sectors and direct financial incentives for energy and water conservation and for reduction of discharges into the environment can have a much more direct impact. The removal of price subsidies and the incorporation of environmental costs are two important policy actions for the energy sector. In the water sector, pricing of ground water would be essential to drive water conservation. Furthermore, environmental levies for waste water discharges, as well as for air emissions and solid waste disposal, will improve the cost effectiveness of Cleaner Production and thereby drive the financing of Cleaner Production. At a more general level, trade and investment liberalisation also have proven to be helpful for Cleaner Production and energy conservation, as have investment incentives.

(5) Focus on the Service Sector(Sixth International High-level Seminar on Cleaner Production)

Experts noted that much of the work done to date in Cleaner Production has focused on the manufacturing side of existing production systems. This focus should be expanded to include the service sector. Acquisition of goods as a measure of affluence is replaced by measures of wellbeing based on continuous satisfaction of changing expectations for quality, utility and performance;

The Future: Refreshing the Vision and the Practice

In general terms, there appear to be two major challenges with respect to improving Cleaner Production programmes and policies: 1) enhancing the innovativeness of Cleaner Production programmes in terms of technologies considered, as well as methodologies and instruments used and partnerships established; and 2) broadening and deepening the involvement of non-environmental governmental and non-governmental stakeholders in the design, implementation and management of Cleaner Production programmes (René 2001).

We cannot rely on gradual technological improvements to achieve Cleaner Production. We need quantum leaps, which only innovative technologies can make possible. The primary challenge is to move from rhetoric to implementation. Awareness is important, but it is not the same as taking action. Governments can and must play a leading role in establishing enabling legislative and economic frameworks for Cleaner Production and Sustainable Consumption. Companies should widely apply existing management tools; they should also integrate the environmental dimension into overall decision-making. Consumers, whether they are companies or individuals, need to understand that Cleaner Production and Sustainable Consumption are two sides of the same coin.

Integrating CP into Sustainability Strategies

Since the work of the Brundtland Commission and the subsequent 1992 UN Conference on Environment and Development, the concept of sustainability has been enshrined as the global vision of a healthy future. The past few decades have witnessed significant advances in identifying and understanding a wide range of environmental problems: climate change, global pollution, habitat loss and over-population. The task ahead will require less attention to problem characterization, and more attention to solution development and facilitation. Cleaner production has been a conceptual bridge connecting industrialization and sustainability.It would be useful to consider three rather conventional environmental objectives that will require progress over the next decade: detoxification, dematerialization and decarbonization (Ken 2001).

The UN's international work programme on sustainable consumption and production patterns, adopted in 1995 by the third session of the Commission for Sustainable Development (CSD), defined sustainable consumption as "the use of goods and services that respond to basic needs and bring a better quality of life, while minimizing the use of natural resources, toxic materials and emissions of waste and pollutants over the life cycle, so as not to jeopardize the needs of future generations" (Third Session of the Commission for Sustainable Development 1995). Cleaner Production leads to increases in resource efficiency and deceases in waste generation, reducing the environmental burdens of industrial production. However, steady growth in consumption required increases in the aggregate level of production that could wipe out the impacts of these process improvements. It makes little sense to promote Cleaner Production without seeking to reduce consumption levels, particularly in the wealthier economies. In industrialized countries it means a shift from an emphasis on quantity to one of quality in consumption. The most prevalent consumption approach is demand management, whose success depends on a feedback mechanism that is information rich and culturally attuned. Product innovations such as multi-functionality, extended life products, dematerialization and even e-materialization can all work to make consumption more sustainable (World Business Council for Sustainable Development).

To move from words to action is our primary challenge. The root causes of global environmental degradation lie in current unsustainable production and consumption patterns. The challenge before us is to reorient these unsustainable patterns by promoting a life cycle economy that incorporates a "Cleaner Production" strategy.

ECONOMIC-CUM-ENVIRONMENTAL SUSTAINABILITY

Even the unsatisfying status of environmental problem-solving has, nevertheless, fostered the emergence environmental politics which is dubbed primarily by the catchword sustainability. "The sustainability approach envisions a complex web of human and natural systems interactions and linkages, without starting or end point. This goes well beyond the much more constricted and artificial focus characteristic of earlier thinking and policy formulation that treated air, water, land, and other pollutants separately .

Efforts to achieve a transition to sustainability have been primarily driven by concerns about the ability of the natural carrying capacity of the earth to support the growing impact of human activities. However, sustainability means more than the environment alone. Social and economic sustainability are also at a stake. The success of sustainability-oriented transition management depends strongly on the interactions and performance of the involved players.

Much of the effort to find a pathway towards a sustainability transition has been focused on product and process innovations. Product and process innovations have been and still are important in addressing the problems generated by current production and consumption patterns. These innovations have led to substantial improvements in environmental efficiency (Geels 2005). Energy-efficient and cleaner technologies also are examples of such innovations. These system optimizations have led us – and can continue to do so – to substantial improvements in eco-efficiency. And a societal transition is the result of co-evolution between technological, economical, ecological, cultural and institutional developments on different scale levels (Rotmans, Kemp et al 2000).

But the politics of sustainability brings in much more complexity in environmental policy. Degradation in most developed countries has accelerated as a result of continuing population and economic growth to such an alarming degree that if present practices continue there will be few natural resources left within only a few decades. Continuing economic development in many developed countries, which has been dependent on gross exploitation of resources with little regard for environmental effects, it would soon be no longer sustainable if present practices continue, resulting in widespread economic chaos. Therefore, Economic Policies for Sustained Development is an important first step in a strategy moving towards economic-cum-environmental sustainability. At the academic level there has been a parallel development, with a very high rate of growth in the number of professionals, articles and even journals dedicated to environmental economics. These include changes in policies on national accounting, for example, use green calculation system, allocation of resources by sectors and geography, pricing / taxing / investment/ trade policies, social welfare policy (especially as related to persistent poverty), population growth and distribution, and the development of appropriate industry for job generation. While 'externalities' is one foundation of environmental economics. Economists have often been accused by the 'true environmentalists' from other disciplines of being locked into a subject which is basically structured to deny the existence of environmental problems. Our vocabulary seems to confirm this – environmental problems are something 'external' to the market. Nothing could be more misleading: the concepts of externality, market failure and non-convexities of the production and consumption sets have been at the very heart of general equilibrium theory as long as it has been seriously studied. In other words, and often misunderstood by non-economists, 'externalities' do not mean that environmental consequences of economic activity are external to economic theory. Environmental and ecological elements can be easily integrated into externality theory to make it more 'realistic'(Crocker and Tschirhart 1992).

Much work on externalities can go further than is presently done. If we as economists want to influence policy in the direction of price-based solutions, then we should not stop at qualitative suggestions, but also offer real numbers. Concrete suggestions on price corrections may, however, stimulate the debate on how far from optimality our current prices are. For instance, in the ecotaxation (double dividend) debate, voluminous insights from the theoretical literature are offered, but very few concrete empirical findings. Empirical work in environmental economics should not only try to realize improved assessments of external costs, but also analyze policy options from a perspective that takes all externalities into account simultaneously. Empirical work in environmental economics should not only try to realize improved assessments of external costs, but also analyze policy options from a perspective that takes all externalities into account simultaneously. In road transport, externalities are related to congestion, accidents, emissions of various pollutants and noise. These externalities are not independent. But most studies have adopted a number of too restrictive assumptions, focusing on (carbon) tax instruments only, ignoring variations in marginal costs of control, and not paying attention to monitoring and administration costs. And a great deal of research has already been undertaken on growth theory (exogenous or endogenous). However, almost none of it has been applied or tested empirically. Growth analysis should pay more attention to historical and empirical analysis, to specific aspects of technological change, and to evolutionary theories.

More generally, environmental economists may follow two paths: (1) developing new methods and testing hypotheses, allowing for some degree of abstraction and isolation from real policy making; and (2) studying more practical, integrative, complex issues related to actual policy making, which could generate relevant conceptual issues that are presently overlooked. Many fundamental questions remain unanswered. Environmental economics must be based on sound economic theory but not on a narrow conception of the economy. It must broaden its perspectives from an exclusive attention to efficiency, and integrate various criteria into analyses, including sustainability, equity, and social and political acceptability. The complexities of real societies must be brought into the analysis.

So the objective was to obtain the country's overviews on how to proceed to achieve sustained development. The only truly sustainable development is economic-cure-environmental (El-c-E2) development (ADB 1988,1990; Harvey 1990). The economic decision makers, are now on the threshold of the realization that the only truly sustained development is El-c-E2 development. These decision makers are now more or less ready to modify their approach to economic development planning to accommodate environmental protection needs, their influence could well be the difference needed to solve the country's environmental degradation problems.

The strategy involved the following steps:

1. Evaluation of these in terms of how the economic policies used by the economic decision makers can feasibly be modified to achieve the changeover from essentially economic-oriented development to El-c-E2 development.
2. Preparation of conclusions or guidelines by which the country's continuing development planning should encourage sustainability, including the optimal roles of the country agencies (public and private, including non-governmental organisations) and of the IAAs. Planning strategy to modify conventional El-oriented planning into El-cum-E2/sustained development planning, (which these have proved promising.
3. The operating procedures of the National Economic Planning Agencies (NEcPAs) need to be modified to change from 'economic focusing' to comprehensive El-c-E2 evaluation, so that these agencies may be effectively utilized. This should include strengthening of skills in integrated/quantified environmental analysis, effective use of environmental impact analysis, environmental standards, industrial waste permit systems, environmental monitoring, coordination with other government agencies, enforcement of appropriate regulations, and other basic measures which can be used under NEcPAs leadership for protecting environment within the E1--c-E2 context.

The transition to sustainability is a long-term journey that must be addressed now. Envisioning a sustainable society based on defined parameters and designing strategies and identifying paths towards that end is the approach that serves the sustainability goal.

While it is widely known that sustainable development is the only sound and viable pathway for humankind's future, its attainment remains elusive despite intensive efforts and some successes. The current industrial society approach based on product and process innovation in a variety of fields is not providing the expected results in addressing this important issue.

In an attempt to carry out this unavoidable task, we need an integral and dynamic innovation system where technology plays a key role in fulfilling societal functions. Innovation systems bridge the gap between knowledge and its intrinsic technological possibilities and at the same time match these technological possibilities with market needs.

Since technology management is the key to propelling effective innovation towards sustainability, we can propose a technology transition management through the interaction of technology push, demand pull and institutional design. Practical application of the results of cooperation with industry, such as the reduction of environmental pressure and an increase in resource productivity, contribute to the sustainability goal from an environmental perspective.

CONCLUSIONS

To form complete, harmony and sustainable development principle and to realize sustainable development, we must change the old mode used by many developed countries that is the one "pollute first, manage later" , which is the industrialization based on depleting resources and destroying environment, and must deal with the relations between economic development, population, resources and environment correctly. This article has introduced the relationship of environment and economy. ecological and environmental situation and economic development level follow Environmental Kuznets Curves which presents the shape of inverted U and economy and environment can realize the positive interaction new idea in circular economy which reveals essentially the internal relations of environment and economy. Environmental Kuznets Curves has proved by historical track of developed country and the circular economy model is also exploring and practicing by many countries. Circular economy is the optimal development model which can realize the interaction of economy and environment. The circular economy theories enrich the connotation of Environmental Kuznets Curves, and the practices of circular economy make the Environmental Kuznets Curves more perfect.

The absence in virtually of effective monitoring systems for observing and documenting industrial performance in managing wastes is noticeable. So establishing an appropriate industrial waste permit system and the monitoring to go with it are essential because of the common and effective use of this method of control in the industrialized countries and the availability and practicability of the technology. This article also proposed Ecological management is an embodiment of mechanism innovation and management innovation in the industry which is taking a new road to industrialization. Construction and management must embody the principle of ecological management. In the process of construction and operation, company should implement green management and obligation service system based on the international ecological environment management system ISO14000.

Cleaner Production is the cornerstone of Circular economy. If China wants to vigorously build a resource-saving and environmental-friendly society, Cleaner Production should be an essential part of any comprehensive pollution management system, at the enterprise or the national level. This article puts forward some key strategies for promoting Cleaner Production in China, for instance, integrating CP into sustainability strategies, technology innovations, and industrial ecology. Lack of capacity to track and evaluate the effects of CP programmes, and to disseminate this information, still seems to be one of key challenges for CP initiatives. So development of a Cleaner Production indicator system and recommendation of technical criteria is essential. Cleaner Production has moved industrial environmental management from the end of the production pipe to the end of the innovation pipe, but has not yet been integrated into the innovation cycle as required for the transition towards a priori clean products and processes.

It can often be heard that sciences and technological innovations have a high and still increasing significance for the welfare and the further development of modern industrial societies as well as for non-industrialised countries. While science and science politics are bouncing about knowledge production and hazard production, environmental politics have become a source of producing challenges not only for conventional public policy but also for the sciences. Beside the impact of industrial pollution and harm to many habitats by development, tourism, recreational activities and agricultural practices, a new industry has emerged which has created new, less polluting industrial processes, with processes of remedy of toxic waste sites, with enhancing of energy efficiency etc. Though in some aspects incipiently developed yet, in many industrial branches, this new facet is growing more importantly as a means of making profit. But also for public debates and the self-consciousness of businesses, industries and sciences, one should not underestimate the importance of that emerging industry.

This article also introduced the seeds for development of a new model that could allow us to quantify progress in sustainability. In the context of environmental policy some methods and instruments have emerged to combine attention both short- and long-term impacts of human activities. Such instruments like Life Cycle Analysis, Technology Assessment or Environmental Auditing have proven successful and valuable not only for the environment but for corporations and businesses alike. In this way not only environmentalists but also industrialists and economists are going to shape new concepts of economic welfare, scientific progress and international competitiveness; most of these concepts communicate with the world wide political and intellectual process of implementing Agenda 21 and its core, a politics of sustainability. Through the introduction of an innovation system that seeks to streamline the current technological paradigm and of a technology transition model, this article has offered a framework for the transition to sustainability. We believe that efforts like economic policies are important to ensure for ourselves and future generations a planet that will provide us with the natural capital essential for our development. While this initiative should be an important step in the journey towards sustainability, more needs to be done in this field. The design of a standardized model that could enable us to quantify progress towards sustainability is crucial for this journey. We need tools that will let us measure the different sustainability initiatives and help us concentrate our efforts and resources on the development of the science and technology that will contribute most to the goal of sustainability.

REFERENCES

A World Bank Policy Research Report. Greening Industry: New Roles for Communities, Markets, and Governments. *Oxford University Press,* 2000. 8~11.

ADB. Guidelines for Integrated Regional Economic-cam-Environmental Development Planning. *Asian Development Bank, Environmental Division, Manila.*1988.

ADB. Progress reports of ADB project of 1989-90 on Economic Policies for Sustained Development. 1990.

Bennett, Martin and Peter James. Sustainable Measures: Evaluation and Reporting of Environmental and Social Performance. Greenleaf, Sheffield (UK), 1999.

Chen shiqing, Cai luojia. The Hypothesis of Environmental Kuznets Curve and its Test in China [J]. *Ecological economy*. 2007, 09, 70-73.

Crocker, T. D. and J. Tschirhart. 'Ecosystems, Externalities and Economics' [J]. *Environmental and Resource Economics*. 1992, 02: 551–567.

Davia f. Bradford, Rebecca schlieckert, Stephen h. shore. The Environmental Kuznets Curve: Exploring A Fresh Specification[R]. *C E Sifo working paper,* 2000, 367.

Erkman, Suren. Industrial Ecology: An Historical View, Journal of Cleaner Production, 1997, 5, 1-2: 1-10.

Fan Jin, Hu Hanhui. Studies and Applications of Environmental Kuznets Curve(EKC) [J]. *Mathematics in practice and theory,* 2002, 06: 944-951.

Frosch, Robert A. and Nicholas E. Gallopoulos. Strategies for Manufacturing, Scientific American, special issue on Managing Planet Earth. 1989, 261, 03: 94- 102.

Geels F. Technological transitions and system innovations: a co-evolutionary and socio-technical analysis. Edward Elgar Publ, Cheltenham. 2005.

Gertler, Nicholas and John R. Ehrenfeld. Industrial Ecology in Practice: The Evolution of Interdependence at Kalundborg [J]. *Journal of Industrial Ecology,* 1997, 1, 01: 67-79.

Grossman G, Kreuger A. Economic growth and the environment [J]. *Quarterly Journal of Economic*. 1995 ,110 (2) :353~337.

Hammer B. Financing cleaner production. In: Report for promotion of cleaner production policies and practices in selected developing member countries, Seattle, Wash, 2001.

Harvey E Ludwig. Moving toward Economic-cum- Environmental Sustainability in Asian Developing Countries [J]. *The Environmentalist*. 1990, 10, 4.

Hilton , F. G. H. and A. Levinson. Factoring the environmental Kuznets Curve : evidence from automotive emissions [J] . *Journal of Environmental Economics and Management*. 1998, 35 :126 - 141.

Hong Yang, Luan Shengji. Approach on Kuznets' Relation between Environmental Quality and Economic Growth [J]. *Shanghai environmental sciences*. 1999, 03: 112-114.

Huang Tao, Gao Hong, Liang Runxia. Study on the Analysis of Rural Environmental Pollution Problem and Control Measures [J]. *Journal of anhui agricultural sciences*. 2008, 36(21): 9270-9272.

James andreon, Arik levinson. The Simple Analysis of the Environmental Kuznets Curve[J]. *Journal of Public Economic,* 2001, 8.

John F. Jaworski, David E. Minns,Technology innovations and Cleaner Production: possibilities and limitations. *UNEP Industry and Environment January* – June 2001.

Ken Geiser. Integrating CP into sustainability strategies. UNEP Industry and Environment January – June 2001.

Li Huiming, Bu Xinxin. How to Achieve Win-Win in Environment and Economy ——Some Considerations Inspired by the Environmental Kuznets Curve [J]. *Nankai Journal*. 2003, 01: 63-69.

Liang Shuxuan, Sun Hanwen. Industrial Wastewater in China: Pollution and Affecting Factors Analysis [J]. *Environmental science & technology*.2007, 05: 49-53.

List J A. The environmental Kuznets curve: does one size fit all ? [J] *Ecological Economics*, 1999, 31 (3):409 ~ 423.

Liu Zhiqiang, Chen Jiling. Research on present situation and development tendency of atmospheric environmental quality in China [J]. *Electric power environmental protection.* 2007, 01: 26-30.

Mazmanian, D.A. and Kraft, M.E. 'Toward sustainable communities', Transition and Transformations in Environmental Policy, *MIT Press, Cambridge/MA.* 1999.

OECD (in press) Case Studies of Biotechnology and Cleaner Production. OECD Task Force on Biotechnology for Sustainable Industrial Development. OECD, Paris.

Qu Geping. State from The Environmental Kuznets Curve(EKC) [J]. *Journal of Environmental Management College of China.* 2006, 04, 5-7.

Reller, A. 'Silicone – eine vollsysnthetische Materialklasse macht Geschichte(n)', *GAIA* 9, 13–23. 2000.

René Van Berkel. CP and industrial development. UNEP Industry and Environment January – June 2001.

Romon L. Corruption ,pollution ,and Kuznets environment curve [J]. *Journal of Environmental Economics and Management,* 2000 ,40 :137~150.

Rotmans J, Kemp R, van Asselt M. Transitions & transition management, the case of an emission-free energy supply. International Centre for Integrative Studies, Maastricht, The Netherlands.2000.

See the site for the Institute for Local Self-Reliance (www.carbohydrateeconomy.org).

See, for example, the Royal Dutch Shell Group site (www.shell.com/royal-en/).

Selden T. M, Song D. Environmental quality and development : Is there a Kuznets curve for air pollution emission ?[J] *Journal of Environmental Economics and Management.* 1994 ,27 :147~162.

Sixth International High-level Seminar on Cleaner Production 16-17 October 2000, Montreal, Canada Recommendations. UNEP Industry and Environment January – June 2001.

The Environmental Protection Department of People's Republic. 2007 Report on the State Environment in China. *China forum of environment journalists.*

Third Session of the Commission for Sustainable Development (1995).

Torras M , Boyce J K. Income , Inequality , and pollution :a reassessment of the environmental Kuznets curve [J]. *Ecological Economics* ,1998 ,125 :147~160.

UNEP Environment Programme. Eco-Efficiency and Cleaner Production: Charting the Course to Sustainability. World Business Council for Sustainable Development.

Van Berkel, R., E. Willems and M. Lafleur. The Relationship between Cleaner Production and Industrial Ecology [J]. *Journal of Industrial Ecology,* 1997, 1, 01: 51-66.

Wei Xin, Liu Xinliang, Su Yang. Characteristics of Pollution in Rural Habitats and Analysis on Causation [J]. *China development.* 2007, 04: 98-104.

World Business Council for Sustainable Development. Eco-Efficiency and Cleaner Production: Charting the Course to Sustainability.

Xie Zhenhua. Reflections on Cycle Economical Theory and Policy [J] . *Environmental Protection.* 2004, 01, 3-8.

Yang Kai, Ye Mao, Xu Qixin. Environmental Kuznets characteristics of municipal solid waste growth in Shanghai city [J]. *Geographical Research.* 2003, 01: 60-66.

Zhang Kai. Analysis on the Influence of the Circular Economy to Kuznets Curve [J]. *China Development,* 2005, 01: 16-20.

Zhao Xikang, Li Jianmin, Wang Jinying, Zhou Chunqi. An Empirical Test of Environmental Kuznets' Curve in China [J]. Nankai EconomicStudies.2005, 03: 48-49.

Zhou Hongchun. Circular economy in China and recommendations [J]. *Ecological Economy.* 2006, 2:102-114.

In: Environmentalism: Environmental Strategies...
Editors: E. Kahraman and A. Baig, pp.129-139
ISBN: 978-1-60741-634-0

Chapter 5

STUDYING MOTOR DEVELOPMENT: A BIOLOGICAL AND ENVIRONMENTAL PERSPECTIVE

Carl Gabbard*[1], Priscila Caçola[1] and Tatiana Bobbio[2]
[1] Texas A&M University, USA
[2] State University of Campinas, Brazil

ABSTRACT

Motor development research has experienced greater frequency in top-tier science journals over the past 25 years. While some dialogue has been of *environmental factors*, to a much greater extent, discussions of motor development have focused on the biological, psychological, cognitive, or movement aspects of change. Few would disagree that to truly understand the mechanisms and processes associated with change in human form and level of motor behavior, a more comprehensive and integrated approach is needed – one that links biological and environmental theory. One such approach and the focus of this chapter is the *Developmental Systems Perspective*.

INTRODUCTION

Motor development is the study of *change* in motor behavior as influenced by biological and environmental factors (Gabbard, 2008). Change in this context typically means the observation of growth (change in size) and development (change in level of functioning) over time. Over last 25 years, there has been a substantial increase in the presence of motor development research in top tier journals of human development, psychology and neuroscience. This attraction is due in large part to acknowledgement that level of motor development is a critical factor in child behavior. Additional evidence of this emergence is the observation that aspects of motor development are mentioned with increasing frequency in broad-based theoretical treatises within the fields of cognitive psychology, developmental

* Address: Carl Gabbard; TAMU 4243; College Station, TX 77843-4243; (979) 845-1277; Fax-847-8987; E-mail c-gabbard@tamu.edu

neuropsychology, developmental psychobiology, and neuroscience (e.g., Johnson, Spencer, & Schöner, 2008; Michel, 2001; Thelen, Schöner, & Scheier, 2001).

Although some mention is given to environmental factors, to a much greater extent, discussions of motor development have focused on *dynamic systems theory* (e.g., Adolph, 2008; Fischer & Paré-Blagoev, 2000; Lewis, 2000; Schöner & Thelen, 2006; Thelen & Smith, 2006). We wish to underscore that the term of focus here is '*developmental systems*,' not dynamical systems; with the latter being a sub set of the former. Our intent with this chapter is to point out the potential advantages of using a broader perspective (framework) for the study and understanding of motor development - the *Developmental Systems Perspective.*

As a general framework, the *Developmental Systems Perspective* advocates the notion that human development is the product of changing relations between the developing person and his or her changing multilevel environmental contexts (Gottlieb, 2000; Lerner, 2002). This approach recognizes that a person is a dynamic, self-organizing unit consisting of several systems (e.g., neurological, muscular, skeletal, and cognitive) with multilevels within each system (e.g., Lewis, 2000; Thelen, 1995). From this perspective, contemporary theory such as *dynamic systems theory* is a sub set of the more general developmental systems view. Inherent with this approach is recognition of the reciprocal interactions between our biological characteristics and the environment, which produce change in human form and behavior. This reciprocity has done much to close the long debated issue surrounding the idea that human behavior is the product of nature or nurture (de Waal, 1999; Thelen, 1995). Most contemporary scientists view these two forces as fundamentally intertwined and inseparable, and part of a complex developmental system (de Waal, 1999; Gottlieb, 2000).

From the developmental systems perspective, change in human and motor behavior may be described with theories associated with *Environmental Contexts* and *Biological Systems*. What follows is a sample (certainly not an exhaustive list) of theories with ties to the developmental systems perspective. That is, theories that may be used to gain a more comprehensive understanding of the complex story associated with change in motor behavior. Although a multitude of developmental questions may be relevant, arguably, the most common contemporary line of inquiry deals with the explanation of mechanisms and processes associated with the acquisition of motor skills – commonly referred to as the study of *Perception to Action.*

Environmental Contexts

The term environmental context is associated with the circumstances, objects, or conditions by which one is surrounded. Also known as ecological systems in the field of ecological psychology, examples include the home, school, culture, and social influences within (e.g., family, peers, and coaches); all of which have shown to influence motor development. Among the noted ecological models, *Bronfenbrenner's Ecological Systems* theory and *Gibson's Ecological Perspective*, have close relevance to the study of motor development. Also, see *Newell's Constraints Model* in the section titled COMBINING ENVIRONMENTAL AND BIOLOGICAL PERSPECTIVES; this model examines motor development in regard to environmental, task, and biological constraints (influences).

Bronfenbrenner's Ecological Systems Theory

This model represents one of the most extensive frameworks in the ecological systems theory (Bronfenbrenner, 1986, 2000); one that has enormous potential for studying environmental influence on motor development. This perspective emphasizes the broad range of situations and contexts individuals may encounter as described by five distinct systems: the microsystem, mesosystem, exosystem, macrosystem, and chronosystem. In brief, these systems represent the environmental settings and relationship ranging from the home, to the community, and to the culture in which one lives. Aside from the more obvious influence of the home, a relevant example is city government (parks), which is responsible for the quality of play and recreational opportunities, such as youth sports, playgrounds, and swimming facilities. Also included in this framework are sociohistorical contexts. For example, females of today are much more likely to participate in athletic endeavors than they were 20 years ago. As will be discussed in a subsequent section, settings and events within systems represent "affordances"; opportunities for developing and maintaining motor skill.

This theory looks at a child's development within the context of the system of relationships that form his or her environment. The theory defines complex "layers" of the environment, each having a potential effect on a child's development. A more recent update of the model highlights the child's own biology as a primary environmental agent. The interaction between factors in the child's maturing biology, his immediate family / community environment, and the societal landscape fuels and guides development. Change or conflict in any one layer may cause a ripple effect on other layers. To study a child's development, we must look not only at the child and her immediate environment, but also at the interaction of the larger environmental context. Bronfenbrenner's Process-Person-Context-Time (PPCT) construct has been used effectively to study various relations between the individual and multileveled context. This model accounts for the influences of individuals (person), their interactions with the environment and the responses they provoke from the environment (process), their interactions within immediate settings (context), and changing sociocultural influences on development (time). The results of a dissertation topic internet search via ProQuest, indicated that in 2007, 23 dissertations used Bronfenbrenner's model.

For example, Tudge and colleagues (2003) used the PPCT to focus on the relations between school-relevant activities (including play) of preschool-aged children and teachers' subsequent perception of the children's competence once they had entered school. They observed 3-year-olds' engagement in everyday activities (Process) and their initiation of those activities (Person) over a period covering the equivalent of an entire waking day. Children were drawn from two social classes (Context). The preschool observations were followed by two consecutive years of teacher reports of academic competence following entry into elementary school (Time). Of more direct relevance to the movement domain and the suggestion of models like Bronfenbrenner's, Salmon and Timperio (2007) conducted a review of the literature regarding the environmental effects on children's physical activity. Their findings were that several contextual factors have been reported. For example, safety concerns (e.g., road safety, crime and concerns about strangers), social interaction (e.g., child - peers, neighborhood relationships, other children live in neighborhood close by), and urban design (e.g., connectivity of streets, access and availability of public open spaces and sports facilities). The authors went on to state that unfortunately, no single report has been published using a broad-based ecological model (such as Bronfenbrenner's model). Furthermore, they recommended that "there is a need for conceptual models that take such complexities into account, and need for multilevel study designs that incorporate individual-level influences, proximal social influences and influences within the broader environment in order to better understand physical activity behavior" (p. 196). Along similar lines, Vrazel, Saunders and Wilcox (2008) gave a recommendation after reviewing environmental effects on women's physical activity. That is, they proposed the need for a social-environmental framework such as Bronfenbrenner's model for studying women's physical activity behavior.

Gibson's Ecological Perspective

This popular view provides insight into an important question in motor development: How do individuals perceive and act on information in the environment? Proposed by Eleanor and James Gibson (1979, 1988, 2001), this view contends that infants can directly perceive information in the environment and act with a reasonable response. This is in contrast to the traditional constructivist notion suggesting that past experience is critical in order to act on information in a meaningful manner. In the Gibsonian perspective, perceiving is experiencing. The infant is an active explorer in this process where perception and motor action are coupled. That is, we cannot study perception independent of movement. The environment, cast in the ecological settings noted in Bronfenbrenner's systems, for example, provides affordances that invite and challenge the child to perceive and act on information. In addition to the more obvious set of affordances such as toys, materials, apparatus, and availability of space, stimulation and nurturing by parents (and others) provide the additional component of events. The notion of affordances emphasizes that there is an ecological fit between the individual and the situation. The study of affordances in perception and as agents for change has been given considerable attention in the motor development and ecological psychology literature (e.g., Clark & Uzzell, 2002; Rodrigues, Saraiva, & Gabbard, 2005; Hirose, 2002; Stoffregen, 2000; Zwart, Ledebt, Fong, de Vries, & Savelsbergh, 2005).

An infant's world could be compared with a playground, where each piece of equipment affords a new motor action. Motor planning decisions, especially novel ones, are made based on the affordance available in the environment, level of experience, and the infant's body dimensions (size). In the later stages of crawling, for example, infants begin to perceive and act upon affordances such as coffee tables and small benches, where they can pull up their bodies and stand. It is not rare to see babies using furniture to perform a motor action that otherwise would be impossible. After accomplishing this task, the action of standing up constantly helps them acquire some muscular strength and balance to remain in that position. At this point, the couch that is only a few steps away is another challenge and opportunity for action. Initially, the baby struggles to walk a few steps in order to reach the couch, but when they do so, a whole new world of affordances opens up. According to Berger and Adolph (2007, p. 240), "an infant's ability to detect affordances is especially impressive because the possibilities for action change from week to week".

In one study designed to investigate how and when infants learn to climb stairs, Berger, Theuring, and Adolph (2007) found that infants' access to stairs influenced the acquisition of stair climbing milestones. Infants with a set of stairs in their home crawled, walked, and went up stairs earlier than infants without it. As the authors stated, when given the opportunity, some infants in the sample climbed a full flight of stairs at 6 months of age. Of course, other environmental contexts play a role in acquisition of motor development milestones, such as cultural aspects and family practices. In a previous study, Berger and Adolph (2003) explored the idea that an implement found in a child's environment affords other motor action. For example, by about 10 months, infants are able to use a rake in order to drag a toy that is out of reach; in essence, the rake is an affordance. Most research with infant tool use has focused on hand-held implements; however, more recent work shows that "experienced walking infants exhibit whole body tool use to expand affordances for locomotion" (Berger & Adolph, 2007, p. 243). In this particular study, toddlers were encouraged to cross over narrow and wide bridges, some of which had a handrail available, and some did not. All toddlers in this study attempted to walk over wider bridges, but were more likely to walk over the narrowest one only when a handrail was available. The results showed that children were able to perceive the handrail as an affordance to augment their balance; therefore they could carry out an otherwise impossible task.

The two studies described show that experienced infants can detect affordances for stance as well as for locomotion. They can even detect tools, such as the handrail, which, like other affordances and objects in the environment, can be used to augment balance, thus making the action on other affordances possible. According to Berger and Adolph, infants restructure the relationship between the actor and the environment to facilitate the completion of an action. In another study using the affordance of gap crossing, Zwart, Ledebt, Fong, deVries, and Savelsbergh (2005) concluded that walking experience (rather than the age of the child), was a significant predictor of gap crossing thresholds in young children. The authors noted, "appropriate coupling of action and perception arises through exploration of the environment" (p. 145). Experience, as mentioned before, is crucial in how children perceive and act on affordances.

BIOLOGICAL SYSTEMS THEORY

The three perspectives appropriately grouped as *developmental biodynamics* (Lockman & Thelen, 1993; Thelen, 2000), complement the developmental systems perspective by providing unique insight into the mechanisms and processes involved in the development of coordinated movement. Arguably, comprehensive understanding of the development of coordinated action requires information about the ecological surroundings.

Developmental Biodynamics

One of the most exciting and promising biological approaches to the comprehensive study of motor development emerged in the 1990s. Sparked by advances in neuroscience, biomechanics, and behavioral science, developmental biodynamics represents an interdisciplinary attempt to integrate promising theories and findings related to the behavioral study of perception and action (Bertenthal & Clifton, 1998; Lockman & Thelen, 1993; Thelen, 2000). In essence, the primary attempt is to describe and explain the intimate connection between the brain and body. This approach represents a vast improvement from the more traditional maturational approach, which described change in a descriptive series of developmental milestones. Although it is generally accepted that the control of motor behavior is under the direction of some type of generalized motor program (GMP [Schmidt, 1975; Shea & Wulf, 2005]) what the program actually controls and how control develops has been two of the most active areas of contemporary motor behavior research. This general approach is based on the hypothesis that motor coordination and control emerge from continual and intimate interactions between the nervous system and the periphery, the limbs and body segments (in essence, the brain and body). From this multifaceted approach, three theories have emerged that address the complex processes and mechanisms involved in motor development (perception to action): *coordinative structures, dynamic systems and neuronal group selection theory.* Another theory that has drawn the attention of the motor development community is *Newell's Constraints Model*, which will be considered separately.

Coordinative Structures

In past years, it was generally assumed that the individual muscles were given instructions by the commands produced in the motor program. Bernstein (1967) challenged this notion. He argued there are too many independent operations for the motor program to control them all simultaneously. In more recent years researchers have supported Bernstein's initial concern (e.g., Thelen, 2000; Turvey, 1990; Vereijken, 1991). They find it difficult to consider that the nearly 650 muscles and 100 mobile joints of the body are under the control of the central nervous system in all the possible degrees of freedom (i.e., controlled action of an independent joint movement). Bernstein suggested that motor programs control groupings of muscles instead of individual muscles. These groupings of muscles with associated joints have been termed coordinative structures, dynamic structures, and *synergies.* The commands generated by the motor program are directed toward the specific coordinative structure that is constrained to act as a single functional unit. From this general notion, a new definition of the term coordination has been proposed. It is the process by which an individual constrains, or condenses the available *degrees of freedom* into the smallest number necessary to achieve a goal (Rose & Christina, 2006). Although the actual nature of how the motor program controls movement patterns is not well understood, the existence of coordinative structures has been verified (e.g., Gentilucci, Chieffi, Scarpa, & Casetiello, 1992; Katsumata, 2007; Lee, Bhat, Scholz, & Galloway, 2008; Saling, Alberts, Stelmach, & Bloedel, 1998). The coordinative structures idea has stimulated considerable inquiry in developmental research, much of which as been associated with the next and arguably most published approach – the dynamic systems perspective.

Dynamic Systems Theory

This perspective seeks to provide an understanding of how movement and control emerge and unfold developmentally. Based on highly complex principles from theoretical physics, mathematics, and ecological psychology, this theory proposes that qualitative changes in motor behavior emerge out of the naturally developing dynamic properties of the motor system and coordinative structures (Kugler, Kelson, & Turvey, 1982). Thelen and Smith (2006) describe the following features of dynamic systems:

1. Movement appears to emerge from *self-organizing properties of* the body. Individuals are composed of several complex and cooperative systems (e.g., perceptual, postural, muscular, and skeletal), with each developing at different rates; thus the intricacy *of* coordinated movement. Even the simplest skill requires a cooperative effort, with each system providing critical input. For example, muscular strength of the legs and postural control are critical controllers for walking. Supposedly, there is a rate controller for each skill that organizes the system or systems needed to execute the task.
2. Another assumption is that movement is determined not only by muscular forces, but also by mechanical interactions (e.g., body segments and joint reaction forces). The contribution of these forces may be influenced by speed of movement, body position, length and mass of body segments, and the intentions of the movement. Research in this area seeks to examine the developmental characteristics associated with the generation and apportionment of forces and the timing-based control system.
3. With regard to the question of continuity or discontinuity, the dynamic systems perspective suggests that in the transition from old movement patterns to new ones, referred to as *a phase shift,* disruptions (discontinuities) occur in performance. The major challenge is to account for these characteristics from processes that are inherently continuous.
4. In addition to the biological contexts of dynamic systems, the nature of the environment and demands of the task influence development. Dynamic systems theory suggests that individual constraints can act as rate limiters in the development of motor skills (see discussion of constraints).

Neuronal Group Selection Theory

The dynamic systems perspective provides a glimpse into self-organizing properties of developing and mature motor systems, but it does not typically identify specific underlying neural mechanisms. The study of *neuronal group selection* (Edelman, 1992; Sporns & Edelman, 1993) focuses on how changes in brain circuitry that control synergies match to developmental changes in the musculoskeletal system. During the early stages of development, neuronal circuits are not precisely wired to execute specific skills; instead the brain contains variant circuits (structural variability) with dynamic properties. That is, those circuits selectively form *neuronal groups, which* account for the brain's organization of synergies as functional units of motor control (coordinative structures). One of its merits is that it accounts for the spontaneous adaptability of coordinative structure in response to biomechanical and environmental changes. Therefore, as with dynamic systems theory, neuronal group selection supports the environmental influence on motor development. Refer to Hadders-Algra (2000) and Jouen and Molina (2005) for additional information on this attractive theory.

COMBINING ENVIRONMENTAL AND BIOLOGICAL PERSPECTIVES

Newell's (Constraints) Model (Newell, 1986) offers an excellent framework for the study of lifelong motor development. It combines both the biological and ecological systems perspective. This is applied by describing the constraints to behavior in reference to the individual, the task to be performed, and the environment in which it is to be executed. In this context, the term constraint refers to factors that either facilitate or restrict development. Underscoring this view is the perspective that new motor behaviors emerge as a result of changing individual (organismic), environmental, and task constraints. In essence, this model supports the developmental systems perspective.

Individual constraints referred to originally as organismic factors, can be divided into two categories: structural and functional constraints. For example, weight, height, and reach may be structural constraints, while speed, coordination, postural stability, and strength are considered functional factors. It is not difficult to understand that, depending on the task, we could be limited to some extent by strength, flexibility, and balance.

Environmental constraints can be related to the physical environment or sociocultural factors. This may include gravity (terrain), surface, space, temperature, and characteristics of the home. For example, the space and terrain that an infant has available to move in is a constraint on the development of locomotion.

Task constraints are broadly grouped into categories of task goal, task rules, and equipment / materials used with the task. Closely associated with the first two constraints (goal and rules) are the cognitive demands of the activity. In regard to equipment - a young child may not be successful with a normal size bat or racquet, but after modification more success or a better movement pattern may emerge. The same concept applies to the infant learning to reach and grasp, in regard to object size. Obviously, there are more specific constraints then those mentioned. Certainly, one can think of numerous situations across all age groups. And, that is the merit of the constraints model. It can be used for evaluating performance and influence the type of instructional strategies used.

Numerous studies have been designed using the Constraints model, with most focusing on the biological constraints associated with motor development. Obviously, studies of affordances, as described with the Gibson's model, could be viewed from the Constraints perspective. The following is an example of a study that used the Constraints model. Summers, Larkin, and Dewey (2008) evaluated activities of daily living of younger (5 – 7 years of age) and older (8 – 9 years of age) children with and without Developmental Coordination Disorder (DCD). The researcher's design incorporated the competence (constraints) of the child, context of the environment (family), and specific self-maintenance tasks in which the child was engaged. The researchers found that children with DCD had more difficulty with dressing, personal hygiene, and eating skills. Difficulties with biological constraints, postural control and fine-motor skill, were reported to contribute to poorer performance. Furthermore, it was clear that the family was important in regard to coping strategies that were created to lessen the impact of DCD on participation in activities of daily living. The findings of this study verified that ecological models such as Newell's may be used effectively to study motor development as influenced by person-environment-task interactions.

CONCLUSION

A commonality among all the approaches described is the suggestion that the brain and body are not pre-wired for skilled movements; rather, they have amazing adaptable *self-organizing properties* that adjust for biological and environmental contexts. In this dynamic process, affordance, exploration and attention to the environment are critical factors. Thus, if the student, teacher, or researcher wishes to examine motor development from a broader perspective, the *Developmental Systems* framework can be most accommodating.

REFERENCES

Adolph, K. (2008). Motor and physical development: Locomotion. *Encyclopedia of Infant and Early Childhood Development*, 359-373.

Berger, S. E., & Adolph, K. E. (2003). Infants use handrails as tools in a locomotor task. *Developmental Psychology, 39,* 594-605.

Berger, S. E., Theuring, C., & Adolph, K. E. (2007). How and when infants learn to climb stairs. *Infant Behavior & Development, 30,* 36-49.

Berger, S. E., & Adolph, K. E. (2007). Learning and development in infant locomotion. In Infants use handrails as tools in a locomotor task. In C. von Hofsten & K. Rosander (Eds.). *From action to cognition* (pp. 237-255). Amsterdam, The Netherlands: Elsevier.

Bernstein, N. (1967). *The coordination and regulation of movements.* Peragmon Press: London.

Bertenhal, B., & Clifton, R. (1998). Perception to action, In L. W. Damon (Ed.), *Handbook of child psychology: Vol 2. Cognition, perception and language* (pp. 51-102). John Wiley & Sons: New York.

Bronfenbrenner, U. (1986). Ecology of the family as a context for human development: Research perspectives. *Developmental Psychology, 22,* 723-742.

Bronfenbrenner , U. (2000). Ecological theory. In A. Kazdin (Ed.), *Encyclopedia of psychology.* American Psychological Association and Oxford University Press: New York.

Clark, C. (1995). On becoming skillful: Patterns and constraints. *Research Quarterly for Exercise and Sport, 66,* 173-183.

Clark, C., & Uzzell, D. L. (2002). The affordances of the home, neighborhood, school and town center for adolescents. *Journal of Environmental Psychology, 22,* 95-108.

de Waal, F. B. M. (1999). The end of nature versus nurture. *Scientific American, December, 94,* 99.

Edleman, G. M. (1989). *The remembered present.* Basic Book: New York.

Fischer, K., & Pare -Blagoev, J. (2000). From individual differences to dynamic pathways of Development. *Child Development, 71, 850*-853.

Gabbard, C. (2008). *Lifelong Motor Development* (5th ed.). San Francisco, CA: Benjamin Cummings.

Gentilucci, M., Chieffi, S., Scarpa, M., & Casetiello, U. (1992). Temporal coupling between transport and grasp components during prehension movements. *Behavioural Brain Research, 47,* 71-82.

Gibson, E. J. (1988). Exploratory behavior in the development of perceiving, acting, and the acquiring of knowledge. *Annual Review of Psychology, 39,* 1-41.

Gibson, E. J. (2001). Perceiving the affordances: a portrait of two psychologists. Erlbaum: Mahwah, NJ.

Gibson, J. (1979). *An ecological approach to perception.* Houghton Mifflin: Boston.

Gottlieb, G. (2000). Nature and nurture theories. In A. Kazdin (Ed.), *Encyclopedia of Psychology.* Washington, DC, & New York: American Psychological Association and Oxford University Press.

Hadders-Algra, M. (2000). The neuronal group selection theory: An attractive framework to Explain variation in normal motor development. *Developmental Medicine & Child Neurology, 42,* 566-572.

Hirose, N. (2002). An ecological approach to embodiment and cognition. *Cognitive Systems Research, 3,* 289-300.

Johnson, J. S., Spencer, J. P., & Schöner, G. (2008). Moving to higher ground: The dynamic field theory and the dynamics of visual cognition. *New Ideas in Psychology*, 26(2), 227-251.

Jouen, F., & Molina, M. (2005). Exploration of the newborn's manual activity: A window onto early cognitive processes. *Infant Behavior and Development, 28*(3), 227-239.

Katsumata, H. (2007). A functional modulation for timing a movement: A coordinative structure in baseball hitting. *Human Movement Science, 26*(1), 27-47.

Kelso, J. A . (1995). *Dynamic patterns.* MIT Press: Cambridge, MA.

Kugler, P. N., Kelso, J., & Turvey, M. (1982). On the control and coordination of naturally developing systems. In J. A. Kelso, J. E. Clark (Eds.), *The development of movement control and coordination* (pp. 5-78). John Wiley & Sons: New York.

Lee, H. M., Bhat, A., Scholz, J. P., & Galloway, J. C. (2008). Toy-oriented changes during early arm movements IV: Shoulder–elbow coordination. *Infant Behavior and Development, 31*(3), 447-469.

Lerner, R. (2002). *Concepts and theories of human development* (3rd ed). Erlbaum: Mahwah, NJ.

Lewis, M. (2000). The promise of dynamic systems approaches for an integrated account of human development. *Child Development , 71,* 36-43.

Lockman, J. J., Thelen , E. (1993). Developmental biodynamics: Brain, body behavior connections - Special feature. *Child Development , 64,* 953-1190.

Michel, G. F. (2001). The developmental-psychobiological approach to developmental neuropsychology. *Developmental Neuropsychology, 19,* 11- 32.

Newell, K. M. (1986). Constraints on the development of coordination In MG Wade, HT Whiting (Eds.). *Motor development in children: Aspects of coordination and control* (pp. 341-361). Martinus Nijhoff Publishers: Amsterdam.

Rodrigues, L. P., Saraiva, L., & Gabbard, C. (2005). Development and construct validation of an inventory for assessing the home environment for motor development. *Research Quarterly for Exercise and Sport, 76,* 140-148.

Rose, D., & Christina. R. (2006). *Multilevel approach to the study of motor control and Learning* (2nd ed.). San Francisco: Benjamin Cummings.

Saling, M., Alberts, J., Stelmach, G., & Bloedel, J. (1998). Reach-to-grasp movements during obstacle avoidance. *Experimental Brain Research, 118,* 251-258.

Salmon, J., & Timperio, A. (2007). Prevalence, trends and environmental influences on child and youth physical activity. *Medicine and Sport Sciences, 50*, 183-99.

Schmidt, R. A. (1975). A schema theory for discrete motor skill learning. *Psychological Bulletin, 82*, 225-260.

Schoner, G., & Thelen, E. (2006). Using dynamic field theory to rethink infant habituation. *Psychological Review, 113*, 273-299.

Shea, C. H., & Wulf, G. (2005). Schema theory: A critical appraisal and reevaluation. *Journal of Motor Behavior, 37* (2), 85-101.

Sporns, O., & Edleman, G. M. (1993). Solving Berstein's problem: A proposal for the development of coordinated movement by selection. *Child Development, 65*, 960-981.

Stoffregen, T. (2000). Affordances and events. *Ecological Psychology, 12*, 1-28.

Summers, J., Larkin, D., & Dewey, D. (2008). Activities of daily living in children with developmental coordination disorders: dressing, personal hygiene and eating skills. *Human Movement Science, 27*, 215-229.

Thelen, E. (1995). Motor development – A new synthesis. *American Psychologist, 50*, 79-95.

Thelen, E. (2000). Motor development as foundation and future of developmental psychology. *International Journal of Behavioral Development , 24*, 385-397

Thelen, E., Schoner , G., & Scheier, C. (2001). The dynamics of embodiment: a field theory of infant perseverative reaching. *Behavioral and Brain Science, 24*, 1-34.

Thelen, E., & Smith, L. B. (2006). Dynamic development of action and thought. In W. Damon & R. Lerner (Eds.), *Handbook of child psychology* (6th ed.). New York: Wiley.

Tudge, J. R. H., Odero, D. A., Hogan, D. M., & Etz, K. E. (2003). Relations between the everyday activities of preschoolers and their teachers' perceptions of their competence in the first years of school. *Early Childhood Research Quarterly, 18*(1), 42-64.

Turvey, M. T. (1990). Coordination. *American Psychologist , 4*, 938-953.

Vereijken, B. (1991). *The dynamics of skill acquisition.* Unpublished dissertation, Free University, The Netherlands.

Vrazel, J., Saunders, R. P., & Wilcox, S. (2008). An overview and proposed framework of social-environmental influences on the physical-activity behavior of women. *American Journal of Health Promotion, 23*, 2-12.

Whitacre, C., & Shea, C. (2002). The role of practice variability in retention parameter transfer and effector transfer. *Research Quarterly for Exercise and Sport, 73*, 47-57.

Woolacott, M. (1993). Age-related change in posture and movement. *The Journals of Gerontology, 48*, 56-60.

Zwart, R., Ledebt, A., Fong, F. F., de Vries, H., & Savelsbergh, G. J. P. (2005). The affordance of gap crossing in toddlers. *Infant Behavior and Development, 28*(2), 145-154.

In: Environmentalism: Environmental Strategies... ISBN: 978-1-60741-634-0
Editors: E. Kahraman and A. Baig, pp.141-150

Chapter 6

PHYTO-BIOREMEDIATION OF POLLUTED MARINE SEDIMENTS: THE NEED OF A BIO-PHYSICAL APPROACH

***V. Bianchi*[*1], *G. Masciandaro*[2], *B. Ceccanti*[2], *E. Peruzzi*[2] *and R. Iannelli*[1]**

[1] University of Pisa, Department of Civil Engineering, via Gabba 22, 56122 Pisa, Italy;
[2] National Research Council, Institute for the Ecosystem Studies, via Moruzzi 1, 56122 Pisa, Italy

ABSTRACT

A meso-scale pilot plant was set up to test the efficiency of a phyto-bioremediation technique applied to slightly contaminated marine sediments dredged from the port of Livorno (central Italy). Different plants were used in combination (a grass, *Paspalum vaginatum* sp. and a shrub, *Tamarix gallica* sp.) with earthworms (*Eisenia foetida* sp.) to improve the bio-physical properties of the contaminated sediments. The objective was to transform the degraded sediment into a fertile and nutrient-rich matrix, that can be at the same time efficiently phytoremediated. The plant-microorganism-earthworm association is a powerful ecological system working at the same time, at three levels: plants release organic exudates, microorganisms mineralize them and grow so releasing nutrients for plants, and earthworms feed on microorganisms. The experiment was performed in columns at pilot-scale. The sediment was biophysically preconditioned with manure and inert bulking materials (5:1 v/v). This constructed system allowed plants to grow in an initial fine textured and salty medium, unfriendly for plant species. The efficiency of the recreated ecosystem was confirmed by the improvement of physical properties of the sediment. This included both permeability and water content, starting from an initial value of 10^{-5} m sec^{-1} and 60% (asphyxial medium) respectively, and reaching at the end of the experimentation, value of 10^{-3} m sec^{-1} and 40%, respectively. Furthermore, significant stimulation of ecological parameters that drive the bioremediation were obtained, such as nutrients (i.e. carbon, nitrogen, phosphorous), microbial dehydrogenase

* corresponding author: v.bianchi@ing.unipi.it

and β-glucosidase enzymes, and humic extractable carbon. Stepwise regression statistics was used to focus the correlation between the physical and biochemical properties of the reconstructed matrix. The statistical approach confirmed that the presence of organic macro-aggregates in the sediment (e.g. humic substances) allows the permeability of the medium to increase, also stimulating enzyme activities (e.g. dheydrogenase and β-glucosidase) involved in the mineralization of the carbon-cycling and mineral nutrient release.

Keywords: Bioremediation, Phytoremediation, Marine dredged sediments, Bio-physical properties, Eisenia foetida.

INTRODUCTION

Bioremediation and phytoremediation, when properly managed, represent environmentally sound and cost-effective methods for treating soils or other matrices containing both organic chemicals (exhausted biofilters, asphalts, refinery sludge, etc) and metals, by using plants and micro-organisms (Marín et al., 2005; 2006; Tejada et al., 2008). Phyto-bioremediation may then enable an appropriate reuse of remediated matrices, minimise disposal to landfill, whilst providing for adequate protection of human health and the environment (EPA, 2005). Recently, phyto-bioremediation has been evaluated also for the treatment of dredged sediments (Meers et al., 2005; Iannelli et al., 2006; Bianchi et al., 2008). Contamination of sediments from the dredging of rivers or coastal harbours represents a problem of notable importance due to both its environmental impact (Ankley et al., 1989; Giesy and Hoke, 1989; Hartwell et al., 1997; Hatakeyama and Yokoyama, 1997; EPA, 1997; Ho et al., 2002) and in terms of number of interested cases, volume of involved material and economic extent (Mulligan et al., 2001; Lors et al., 2004). Traditional remediation techniques are often economically unacceptable because of the large amount of contaminated sediments (Förstner and Calmano, 1998; Meers et al., 2003). In this scenario, phyto-bioremediation technique could play an important role. Since sediments, and particularly those from a marine environment, have poor hydraulic and agronomic characteristics, their mixing with selected materials is probably the only way to make phyto-bioremediation an effective and successful remediation technique (Bianchi et al., 2008). Furthermore, the possibility to recreate a life-hospitable environment within an unnatural polluted matrix, depends on the recreation of an ecosystem in which different environments interact and cooperate. Hospitable conditions were probably ensured by the addition of good quality organic manure, which, being rich in microorganisms and degradable compounds, supplies metabolic energy, promotes humification, releases nutrients, and induces specific microbial growth. Plant roots release organic exudates, microorganisms mineralize them and grow, so delivering nutrients for plants, and earthworms feed on microorganisms that colonise the polluted matrix. The evolution of the biochemical and physical properties of a polluted matrix guides in the evaluation of the efficiency of the phyto-bioremediation technique. This means that a bio-physical approach is fundamental to detect the evolution of a phyto-bioremediation process.

In this work, an integrated system based on simultaneous plant-microorganism-earthworm action (reconstructed ecosystem) is proposed for the revitalization and restoration of the biological functions of a sterile and heavily degraded geological resource, such as the contaminated dredged sediments coming from Livorno port (central Italy). Although they operate in a perfect closed cycle, plants, microorganisms and earthworms develop specific ecological functions (Bianchi and Ceccanti, 2008, in press). The reconstructed ecosystem confers to the sediment the fundamental ecological functions similar to those occurring in a natural and healthy soil, from a physical and biochemical point of view. Healthy, fertile soil is a mixture of water, air, minerals, and organic matter (NRCS, 2008).

In the attempts to assess both efficiency and evolution of the remediation system toward a natural ecosystem, transforming the sediment into an active and nutrient-rich soil, two classes of marker parameters were selected in this work: those describing the nutrient status and activity of the microbiological components and those describing the evolution of the physical properties of the sediment.

The overall metabolic activity in the system was measured through the enzyme dehydrogenase (DH-ase) and β-glucosidase. In particular, DH-ase is a constitutive enzyme of all micro-biota usually supported by water soluble carbon substrates, while β-glucosidase is an hydrolytic enzyme of the C-cycle responsible of the cellulose degradation (Masciandaro et al., 1997).

A good permeability and aeration in the root-zone contributed to the stimulation of the soil enzyme activity and to the consequent reduction of organic pollutants (i.e. hydrocarbons). In this study, the detection of the sediment water content was performed using a capacitance probe.

In order to determine the most significant regression equation for estimating the bio-physical amelioration of the sediment applying the proposed phyto-bioremediation technique, the physical and biochemical data were analyzed by means of multiple stepwise regression.

The reconstructed ecosystem is expected to be resilient and to evolve spontaneously from an initial technogenic phase towards a natural soil. For this reason, the main object of the study was the bio-physical amelioration of the degraded matrix.

MATERIALS AND METHODS

1. Basic Description of the Mesocosms

The experimentation consisted in columns constructed to be operated at a meso-scale level (mesocosms). The mesocosms were 1 m high with a 31 cm diameter, equipped with a drainage system at the bottom and placed in the greenhouse of the National Research Council, Institute for the Ecosystem Studies (CNR – ISE Pisa, central Italy), sheltered from the rain. Two different types of mesocosms were prepared in triplicate: 1) mesocosms (controls) containing a mix of dredged sediment and inert bulking material from demolition and excavation activity (5:1 v/v) [C]; 2) mesocosms built in the same way as C and planted with *Tamarix gallica* (leguminous bush) and *Paspalum vaginatum* (graminae grass), added with 30 adult earthworms (*Eisenia foetida*, Savigny) [PME]. More details of the meso-scale plant are reported in Bianchi, 2006.

The combination of grass and bush species is proposed to simulate a first colonization of the top 0-20 cm layer of sediments by a salt tolerant graminae with a consistent root apparatus adaptable to extreme conditions, then followed by bush plantation which may colonize deeper layers (> 20 cm), already tested in a meso-scale application (Bianchi et al., 2008b). The top layer of 0-20 cm of the PME mesocosms was mixed with a farmyard manure (10% dry weight basis) which, being rich in microorganisms and slowly available nutrients, recreated a hospitable environment for the plant-earthworm-microorganism association. In fact, this layer represents the biologically-active stratum which was colonized by plant roots, earthworms and microorganisms. The experimentation took place for 18 months, from September 2006 to March 2007.

2. Origin of the Polluted Marine Sediment

In this study sediments slightly contaminated by heavy metals and hydrocarbons from Livorno port (Tuscany Region, central Italy) were analyzed (Bianchi et al., 2006). The main characteristics are reported in table 1. According to a preliminary laboratory study, these sediments turned out to be inadequate for a direct phytoremediation approach, due to their low permeability and low nutrient content (Iannelli et al., 2006). Physical pre-conditioning of the sediment was necessary in order to enhance water-air transport and nutrient circulation inside the pores at the root zone (rhizosphere); physical amelioration better supports microbial life proliferation at the root zone. Furthermore, the high salinity of the sediments prompted the use of two salt-tolerant species (*Paspalum vaginatum* and *Tamarix gallica*). The first was selected because it has already been successfully used for phytoremediation of marine sediments (Hue et al., 2002; Bianchi et al., 2008).

Table 1. Characteristics of the raw marine sediments (S) at the beginning and of the reconstructed plant-microorganism-earthworm ecosystem (PME) after 18 month. Values with the same letters in the rows are not significant at 5 %

	Raw sediment (S)	After remediation (PME)
TOC (%)	1.6 b	4.8 a
TEC (mg kg $^{-1}$)	5200 b	18000 a
TKN (%)	0.5 b	0.9 a
Ptot (mg kg $^{-1}$)	600 b	850 a
Exch-metals (meq g $^{-1}$)	3.8 a	1.4 b
DH-ase (μgINTF $g^{-1}h^{-1}$)	1.2 b	2.8 a
β-glucosidase (μgPNFg^{-1}dw^{-1} h^{-1})	15 b	24 a
k (m sec $^{-1}$)	10^{-6} b	10^{-4} a
Sand (%)	16 b	54 a
Silt (%)	72 a	42 b
Clay (%)	12 a	4 b

3. Analysis

Samplings for chemical and biological analyses were performed every 3 months while water content measurements inside the mesocosms were carried out monthly. The water supply was about 100 ml of tap water per week, for each mesocosm. Variation in the supplied volume was possible depending on seasonal changing. Homogeneous samples (50 g of fresh weight) were taken out for the analysis from both types of mesocosm with a gauge from the biologically active and unsaturated 0-20 cm top layer. Samples from the raw sediment were taken and analysed before the pre-conditioning (S).

Chemical Analyses

Total organic carbon (TOC) and carbon content of the pyrophosphate extracts (total extractable humic carbon, TEC, Masciandaro *et. al.*, 1998) were determined using a RC-412 MULTIPHASE CARBON analyser (Leco, U.S.A.) and by dichromate oxidation (Yeomans and Bremner, 1988), respectively. Total nitrogen (TKN) was determined by flash combustion using a FP-528 PROTEIN/NITROGEN DETERMINATOR (Leco, U.S.A.) and total phosphorus (Ptot) by using the digestion-spectrophotometric method reported by Olsen and Sommers (1982), which measures the phospho-molybdate complex at 720nm. Exchangeable heavy metals were solubilised with Diethylene-triamine pentaacetic acid (DTPA) 0.005 M and CH3COONH4 1 M, pH 6.4 (SISS 2000). For practical purposes, in the statistical elaborations both total and exchangeable heavy metals concentrations have been expressed as the sum of individual meq g^{-1}.

Biochemical Assay

Dehydrogenase activity (DH-ase) was determined by the method of Masciandaro et al. (2000) using 2-p-iodo-3- nitrophenyl 5-phenil tetrazolium chloride (INT) as substrate. The product iodo-nitro-tetrazolium formazan (INTF) of the reaction was determined spectrophotometrically at 490 nm.

The method used to assay hydrolase activity (β–glucosidase) is described by Garcia et al. (1993).

Physical Analysis

The particle size distribution was defined by laser screening (Microtrac, ASVR device). The hydraulic conductivity (*k*) was determined with the steady method, in laboratory (Klute and Dirksen, 1986). Water content was measured using a capacitance probe (DIVINER 2000, Sentek Australia), as a guide for the reintegration of water lost through evapotranspiration and leaching (Bianchi, 2006).

4. Statistics

Statistical analyses of the experimental data were performed using the STATISTICA 6.0 software package (StatSoft inc., USA). All the reported results are means of three replicates. Two-factor analysis of variance (factorial ANOVA) was used to check the statistical significance of observed data differences. The reported significant levels ($p<0.05$) are based on Student's distribution. In order to determine the most significant regression equation for estimating the bio-physic amelioration of the sediment applying the proposed phyto-bioremediation technique, the physical and biochemical data were analyzed by means of a stepwise multiple regression. The stepwise variable selection procedure works essentially in the same manner as multiple regression procedure, except that it can be used as a forward selection to control the entry of variables into the model. In each step of the procedure, variables are entered with the goal of obtaining a model with a set of significant variables.

Results and Discussion

1. Sediment Bio-Physical Amelioration

Table 1 reports the sediment properties at the beginning of the experimentation and after the remediation (18 months), applying the reconstructed plant-microorganism-earthworm ecosystem. The sediments treated with plants and earthworms (PME) showed increases of almost all functional parameters regulating chemical and biogeochemical processes with respect to the raw sediment (S). The reconstructed ecosystem, plant-microorganism-earthworm, probably interacts with a maximum efficiency when sustained by an exogenous source of manure added, but more probably by means of plant root exudates, via processes mediated by the stimulatory action of *Eisenia foetida* (Ma et al., 2003). This is suggested by the values obtained for the carbon forms (TOC and TEC). In particular, a considerable part of plant exudates is transformed into stable carbon (TEC). The enzyme DH-ase increases as a result of microbial population growth, though enzymes activities released by plants (Gramms et al., 1999). Plants enhance microbial populations in number and activity also in extreme degraded environments (Garcia et al., 2005). The proposed sediment treatment led also to an increase in nutrient content (TKN and Ptot; Table 1), which supports plants life in a healthy and fertile soil. Microorganisms produce polymeric humic structures which may affect heavy metal availability through chelation (Garcia-Mina, 2005). *Eisenia foetida* activity increases the microbial populations and the bioavailability of heavy metals in soil for plants, as discussed in other experiments of soil remediation (Wen et al., 2004: Bianchi et al., 2008b). This underlines that earthworms are directly involved at the rhizosphere level as they feed on the microbial biomass of roots, ingest soil and release heavy metals and nutrients available for plant uptake (Ma et al., 2003). Data reported for the exchangeable heavy metals (Table 1) showed a decrease during the experimentation time, indicating the efficiency of the proposed treatment for the remediation of marine sediments. The high values of TEC and the DH-ase activity confirmed the biological activation of the matrix (Table 1). This activation is linked to the improvement of the physical properties of the raw sediments, which seemed to be a silty matrix at the beginning, with percentage of sand, silt and clay of 16%, 72%, 12%, respectively, and becoming a sandy-clay matrix at the end of the experimentation (54% of sand, 42% of silt and 4% of clay) (USDA classification). After 18 months, the hydraulic permeability (*k*, Table 1) had doubled in the rhizosphere. This was due surely to the pre-conditioning of the sediment with inert material, but also to the presence of plant roots and earthworms. The increase of permeability in the rhizosphere (0 – 20 cm layer) surely modified the water content inside the mesocosm mesocosms (figure 1). The untreated mesocosms (mesocosms C) represent a typical asphyxial medium, with a constant level of water content (60%), meaning a saturated medium with stagnant pore water. In the PME mesocosms instead, a better drainage of water was observed in the first 20 cm of depth, in correspondence of the rhizosphere. The better drainage and the presence of the formed organic macro-aggregates in the medium, confirmed by the increase in the hydraulic permeability, surely stimulated microbial activity, as was confirmed by the increase of DH-ase and β-glucosidase values (Table 1). The reconstructed ecosystem displayed characteristics similar to a natural soil with the biological compartments active and balanced. These conditions strengthened the microbial pool, contributing also to physical-structural amelioration of the sediments.

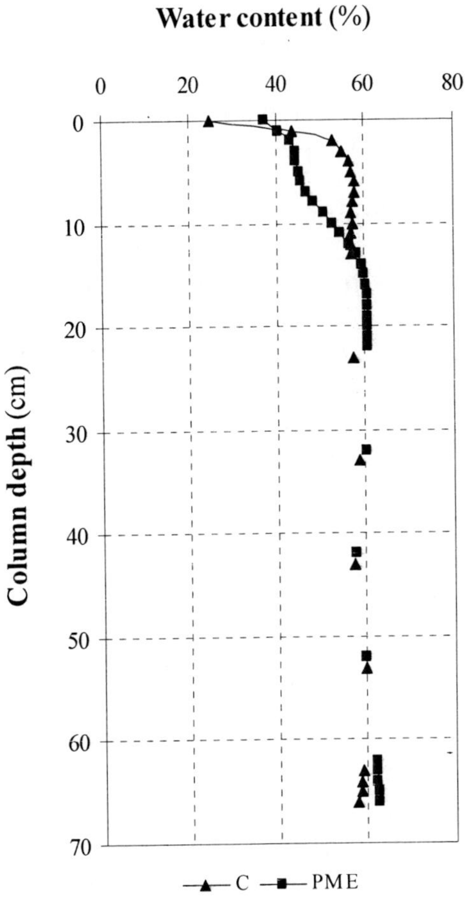

Figure 1. Water content inside the mesocosms. Data registered for the control mesocosms (C) and for the reconstructed plant-microorganism-earthworm ecosystem in the PME mesocosms, after 18 month.

2. Stepwise Forward Regression

Data from all the sampling campaign (three monthly sampling for the chemical and biological analysis, and monthly physical measurements), throughout the experimentation (18 month) were used for the stepwise forward regression. From the results discussed before, the reconstructed ecosystem (plant-microorganism-earthworm) better ensures both sediment amelioration and decontamination. However, once the bioremediation systems have been selected on the basis of their efficiency, emphasis should be given to better interpretation of biogeochemical processes and factors driving the bioremediation and phytoremediation. The stepwise forward regression allows definition of which and how the detected parameters are significantly correlated. Equation 1 reports the most significant correlation found for the physical and biochemical parameters.

$$k = 0.76 \cdot DHase + 0.69 \cdot \beta glu + 0.48 \cdot TEC \quad (1)$$

with $R^2 = 0.76$

The permeability (k) relates positively to the microbial activities (DH-ase and β-glucosidase) and their substrata (humic substances, TEC). This correlation confirmed the link between the physical structure of a soil and biochemical activities, which govern the complex mechanisms of mineralization and degradation of organic compounds in a healthy soil. The better the drainage and aeration of a soil structure, the better is the microbial stimulation inside the medium, detected by oxydoreductase enzymes (Masciandaro et al., 1998). Furthermore, mechanisms for carbon utilization in the proposed experimentation may differ from the usual occurrence in a normal unpolluted soil, as TEC is capable also of binding with heavy metals (Bianchi and Ceccanti, 2008, in press). The reconstructed ecosystem enabled the formation of organic macro-aggregates in the rhizosphere, which improved sediment permeability and drainage, stimulating also the microbial populations (Chamouleau et al., 2001).

CONCLUSION

The reconstructed plant-microorganism-earthworm ecosystem appears to be a valuable and innovative method for the reclamation and restoration of polluted marine sediments. As a matter of fact, the earthworms stimulated microbial activity in the sediment matrix, mixed and aerated the medium, so increasing the permeability, constructed channels for root growth and acidified the medium through excretion, contributing to the bioavailability of the heavy metals. Plant root exudates played a direct effect in the stimulation of microbial communities that grow and degrade organic contaminants in the rhizosphere zone, as well as the possible phytoextraction of available heavy metals. The microorganisms played a direct effect in degrading organic compounds and in restoring the biochemical activation of the sediment. They also produced chelating compounds (e.g. humic substances) which may link with heavy metals. The data obtained throughout the duration of the experiment (18 months) confirmed that the reconstructed ecosystem prompted a spontaneous evolution of the sediment from an initial degraded phase towards a natural and fertile soil, carrying out both a biophysical amelioration of the cleaned matrices and a possible attenuation of mineral-organic contaminants. The statistical elaboration of data underlined the positive and significant correlation between the physical and biochemical parameters, demonstrating the importance of good drainage and aeration for the stimulation of the microbial population, responsible for the mineralization and degradation of the organic compounds, through the enzymes activities.

REFERENCES

Ankley, G.T., Hoke, R.A., Giesy, J., Winger, P.V. (1989). Evaluation of the toxicity of marine sediments and dredge spoils with the Microtox bioassay. *Chemosphere, 18*, 2069–2075.

Bianchi, V., Masciandaro, G., Giraldi, D., Ceccanti, B., Iannelli, R. (2008). Enhanced heavy metal phytoextraction from marine dredged sediments comparing conventional chelating agents (citric acid and EDTA) with humic substances. *Water Air Soil Pollut, 193*, 323-333.

Bianchi, V., Masciandaro, G., Ceccanti, B., Giraldi, D., Iannelli, R. (2008b). Evaluation of phyto-bioremediation techniques applied to dredged marine sediments for their reuse as techno-soils. Third International Meeting on Environmental Biotechnology and Engineering, 3IMEBE, Palma de Mallorca, 21-25 September 2008.

Bianchi, V., Ceccanti, B. (2008). A Three Components System (TRIAS) in the phytoremediation of polluted environmental matrices. *Toxicological & Environmental Chemestry*. In press.

Bianchi., V. (2006). Theoretical and experimental evaluation of phyto-bioremediation techniques applied to dredged sediments. *Master thesis* (Dep. of Civil Engineering, University of Pisa, Italy).

Chamouleau, F., Coulon, D., Girardin, M., Ghoul, M. (2001). Influence of water activity and water content on sugar esters lipase-catalyzed synthesis in organic media. Journal of Molecular Catalysis B: *Enzymatic, 11*, 949–954.

Garcia, C, Roldan, A., Hernandez, T. (2005). Ability of different plant species to promote microbiological processes in semiarid soil. *Geoderma, 124*, 193-202.

Garcia, C., Hernandez, T., Costa, F., Ceccanti, B., Masciandaro, G. (1993). The dehydrogenase activity in a soil as an ecological marker in process of perturbed system regeneration. *XI int. Symp. Environ. Biogeochem.* Salamanca (Spain), 37, 27-30.

Garcia-Mina, J.M. (2005). Stability, solubility and maximum metal binding capacity in metal–humic complexes involving humic substances extracted from peat and organic compost. *Organic Geochemistry, 37*, 1960-1972.

Giesy, J.P., Hoke, R.A. (1990). Freshwater sediment quality criteria: toxicity bioassessment. In: Baudo, R., Giesy, J., Muntau, H. (Eds.), *Sediments: Chemistry and Toxicity of In-place Pollutants*. Lewis Publishers, Boca Raton, FL, pp. 265–348.

Gramss, G., Voigt, K. D. , Kirsche, B. (1999). Oxidoreductase enzymes liberated by plant roots and their effects on soil humic material. *Chemosphere, 38*, 1481-1494.

Hartwell, S.I., Dawson, C.E., Durell, E.Q., Alden, R.W., Adolphson, P.C., Wright, D., Coelho, G.M., Magee, J.A., Ailstock, S., Norman, M. (1997). Correlation of measures of ambient toxicity and fish community diversity in Chesapeake Bay, USA, tributaries – urbanizing watersheds. *Environ.Toxicol. Chem.*, 16, 2556–2567.

Hatakeyama, S., Yokoyama, N. (1997). Correlation between overall pesticide effects monitored by shrimp mortality test and change in macrobenthic fauna in a river. *Ecotoxicol. Environ. Saf., 36*, 148–161.

Ho, K.T., Burgess, R.M., Pelletier, M.C., serbst, J.R., Ryba, S.A., cantwell, M.G., Kuhn, A., raczelowski, P. (2002). An overview of toxicant identification in sediments and dredged materials. *Marine Pollution Bulletin, 44*, 286-293.

Hue, N. V., Campbell, S., Li, Q. X., Lee, C. R., Fong, J. (2002). Reducing salinity and organic contaminants in the Pearl Harbour dredged material using soil amendments and plants. Remediation. New York: Wiley Periodicals, Inc.

Iannelli, R., Giraldi, D., Bianchi, V., Ceccanti, B., Masciandaro, G. (2006). Tecniche di Fitorimediazione applicate al trattamento di sedimenti di dragaggio: analisi preliminari e prove sperimentali di fattibilità. *SIBESA VIII - Italo-Brasilian Simposio on Environmental Engineering.* 17–22 September 2006 Fortaleza, Brazil.

Klute, A., Dirksen, C. (1986). Hydraulic Conductivity and Diffusivity: Laboratory Methods. In: *Methods of soil analysis-Part 1*, Physical and Mineralogical Method, Agronomy Monograph, n. 9, Second Edition, pp. 700-703.

Lors, C., Tiffreau, C., Laboudigue, A. (2004). Effects of bacterial activities on the release of heavy metals from contaminated dredged sediments. *Chemosphere*, *56*, 619-630.

Ma, Y., Dickinson, N.M., Wong, M.H. (2003). Interactions between earthworms, trees, soil nutrition and metal mobility in amended Pb/Zn mine tailings from Guangdong, China. *Soil Biology and Biochemistry*, *35*, 1369-1379.

Marín, J.A., Moreno, J.L., Hernández, T., García, C. (2006). Bioremediation by composting of heavy oil refinery sludge in semiarid conditions. *Biodegradation*, *17*, 251-261.

Marín, J.A., Hernández, JT., García, C. (2005). Bioremediation of oil refinery sludge by landfarming in semiarid conditions: influence on soil microbial activity. *Environmental Research*, *98*, 185-195.

Masciandaro, G., Ceccanti, B., Ronchi, V. and Bauer, C. (2000). Kinetic parameter of dehydrogenase in the assessment of the response of soil to vermicompost and inorganic fertilisers. *Biology and Fertility of Soils*, *32*, 479-483.

Masciandaro, G.,. Ceccanti, B, and Gallardo-Lancho, J.F. (1998). Organic matter properties in cultivated versus set aside arable soils. *Agr. Ecosys. Environ.*, *67*, 267-274.

Masciandaro, G., Ceccanti, B., Garcia, C., 1997. Soil agro-ecological management: fertirrigation and vermicompost treatments. *Biores. Technol.*, *59*, 199-206.

Meers, E., Lamsal, S., Vervaeke, P., Hopgood, M., Lust, N., Tack, F. M. G., 2005. Availability of heavy metals for uptake by Salix viminalis on a moderately contaminated dredged sediment disposal site. *Environmental Pollution*, *137*, 354–364.

Meers, E., Vervaeke, P., Tack, F.M.G., Lust, N., Verloo, M., Lesage, E. (2003). Field trial experiment: phytoremediation with Salix sp. on a dredged sediment disposal site in Flanders, Belgium. *Remediation Journal*, *13*, 87-97.

Mulligan, C.N., Yong, R.N., Gibbs, B.F. (2001). An evaluation of technologies for the heavy metal remediation of dredged sediments. *J. Hazard. Mater.*, *85*, 145-163.

Olsen, S.R., Sommers, L.E. (1982). Phosphorus: In: Page A.L., Miller, R.H., Keeney, D.R. (Eds.). *Methods of Soil Analyses.* Part 2. Chemical and Microbiological Properties. Second Edition. SSSA, Madison, Wisconsin, USA, pp. 403-430.

SISS (Società Italiana Scienza del Suolo) (2000). *Metodi di Analisi Chimica del Suolo.* Franco Angeli ed., cap. XII.

Tejada, M., Gonzalez, JL., Hernández, M.T., García, C. (2008). Application of different organic amendments in a gasoline contaminated soil: Effect on soil microbial properties. *Bioresource Technology*, *99*, 2872-2880.

U.S. Environmental Protection Agency (2005). *Soil bioremediation*, EPA 589/005.

U.S. Environmental Protection Agency (1997). *The Incidence and Severity of Sediment Contamination in Surface Waters of the United States*, EPA 823/R-97/006.

U.S. Natural Resources Conservation Services (2008). Last access: December 2008. Available on line: http://www.nrcs.usda.gov/.

Yeomans, J.C., Bremner, J.M. (1988). A rapid and precise method for routine determination of organic carbon in soil. *Comm. Soil Sci. Plant Anal.*, *19*, 1467-1476.

Wen, B., Hu, X., Liu, Y., Wang, W., Feng, M, Shan, X. (2004). The role of earthworms (*Eisenia fetida*) in influencing bioavailability of heavy metals in soils. *Soil Biology and Fertility*, *40*, 181-187.

In: Environmentalism: Environmental Strategies...
Editors: E. Kahraman and A. Baig, pp.151-167

ISBN: 978-1-60741-634-0

Chapter 7

EXAMINATION OF UNIVERSITY STUDENTS' ATTITUDES TOWARDS ENVIRONMENTAL PROBLEMS: A TURKISH EXAMPLES

***Sibel Erkal**[*1] **and Hande Şahin**[2]*

[1] Associate Professor, Department of Family and Consumer Sciences, Hacettepe University, Ankara/Turkey

[2] Research Assistant, Department of Family and Consumer Sciences, Gazi University, Ankara /Turkey

ABSTRACT

This chapter was designed and conducted to determine the attitudes of university students towards the environment and environmental problems.

In the scope of this descriptive study, a questionnaire form was prepared by using an environmental attitude scale to collect study data. This scale was applied between 30 September 2008 and– 8 October 2008 to students in the Department of Family and Consumer Sciences, Faculty of Industrial Arts Education, Gazi University (n: 358). Three hundred students completed in and returned the questionnaire. The SPSS 13.0 statistical program was used to study data. Mean value, standard deviation, T-test, Levene test, Tamhane test and LSD test were used in the analysis.

We found out that there was no statistically meaningful difference between the mean attitude scores of groups when controlled for variables of age, sex, number of siblings, any environment education taken before university, any out-of-school environment training, and the responsibility of individuals, state and conservationist institutions in the solution of environmental problems. However, students' opinions about how they should be assessed in the ecology class differed significantly in terms grade and income variables. Students listed air and water pollution (42.0%), population growth and urbanization (39.3%), and deforestation (18.6%) as the three most important environmental problems. Considering students' total "Environmental Attitude Scale" score, the highest mean attitude score was given to the "air, water and soil are

* Corresponding Author: Sibel Erkal, Associate Professor, Department of Family and Consumer Science, Hacettepe University, Sıhhıye Campus Ankara/Turkey; Tel: +90 312 3051667/133; Fax: +90 312 3051095;
E-mail: erkalsibel@hotmail.com

inexhaustible resources" item while the lowest mean score was given to the "thinning ozone layer threatens all humanity" item.

Keywords: Environment, environmental problems, environmental attitude scale, university students.

INTRODUCTION

Rapid developments in science and technology improve the life quality of individuals on one hand but, on the other hand, destroy many things and change the structure of many others in their environment [Yılmaz et al. 2002]. The environment is a medium where in people perform their social, biological and chemical activities [Dastan 1999]. Environment is not only a huge area in which we keep on living, but also a huge ecosystem in which millions of creatures live [Yücel, 2004]. In other words, the environment is the external medium where all living things interact with others throughout their whole life [Ertürk 1996].

A rapid and unbalanced increase in world population, unplanned settlements, harmful wastes, rapid decrease in the number of animal and plant species and green areas, heavy traffic, noise, unconscious energy consumption, global warming, increase in natural threats and other incidents that upset the balance of nature and society are the main reasons behind the economic, technological, ecological and sociological problems we face today. Air, water and Soil pollution is increasing [Yıldız, Sipahioğlu, Yılmaz 2005]. A diversity of environmental problems now concerns our whole world. Societes are being affected by major environmental problems as globalization continues, Environmental problems may expand and affect human beings and all living species [Özden,2008].

Global environmental hazards to human health include climate change, stratospheric ozone depletion, loss of biodiversity, changes in hydrological systems and the supplies of freshwater, land degradation and stresses on food-producing systems (WHO,2008) (Figure 1).

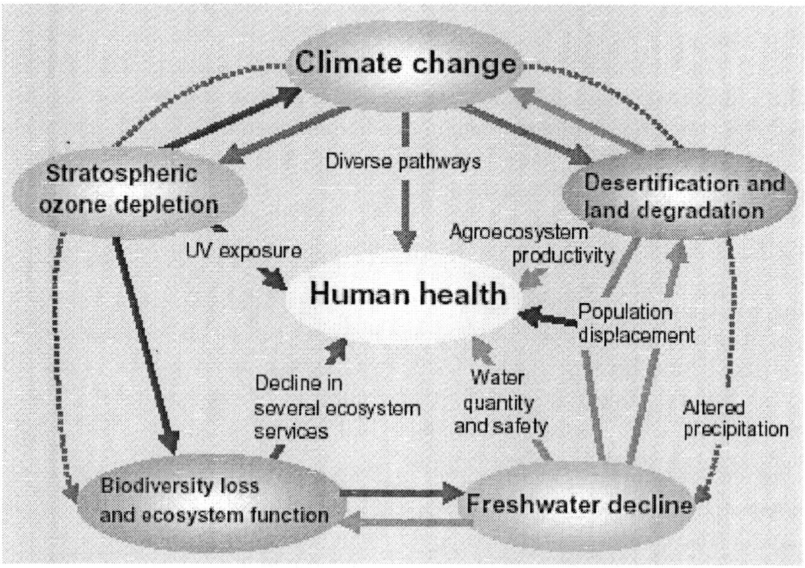

Figure 1. Global environmental change and human health (WHO,2008)

According to the Turkish Constitution: Everyone has the rigt to live in a healthy and balanced environment. Protection of environmental health, prevention of environmental pollutions and development of the environment are the state's and every citizen's duty (Şafak et al.,2008)

Environmental problems that were detected as small-scale local issues in the mid twentieth century have considerably expanded and become global issues today. Local and global measures are taken against environmental problems and national and international meetings are organized to discuss this issue. These efforts have triggered development and adoption of managerial, legal, economic and technological measures against environmental problems [Özdemir et al. 2004, 117].

Determining the awareness is the first step in the solution model of environmental problems. This model covers:

1) Eliciting the awareness for environment and the affecting factors,
2) Providing a real sensitivity to environment by eligible education systems,
3) Accessing the solved environmental problems (Fig.2).

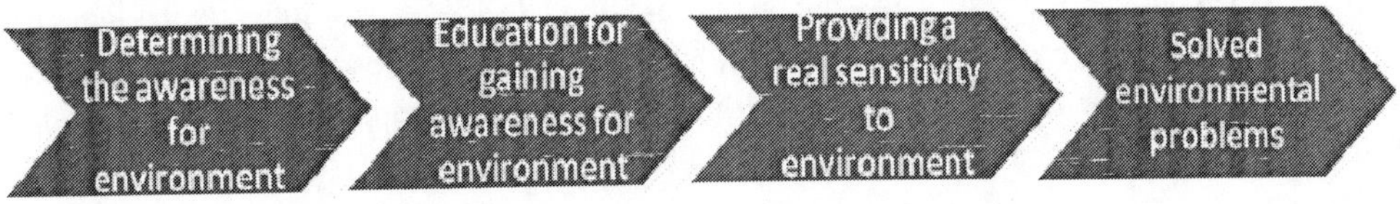

Figure 2.Stages of Soving Environmental Problems (Cankurt et al., 2008)

In relation to environment, the first comprehensive arrangements, concerning International cooperation have been considered in the early 1970s. In 1972 in Stockholm "United Nations Human Environment Declaration", the first global environment of many countries whose socio- economic structures and development levels differ from one another, was accepted[Şafak et al.,2008]. The World Commission on Environment and development published the Brundtland Report (Our Common Future) helping change nd perceiving environment problems as global problems [TÇSV, 1991].

The aim of efforts exerted to protect, develop and improve the environment is to provide individuals with a healthier and safer world to live in. This aim can be achieved by individuals themselves since producing successful ecological results is based on human factor. This is possible by improving human resources, which can be achieved by providing people with required knowledge and skills [Kabaş 2004, 2].

Although until recent years, many resources have been considered unlimited, in the twentieth century people began to be aware of the fact that our environment is subject to serious threats. Environmental consciousness is an important condition for the studies aiming at protecting the environment and preventing pollution; however, it is not adequate by itself. It necessitates participation of all the community. It is difficult to solve the problems unless participation of the individual is provided and wrong habits are changed. A person who feels responsible for his environment chooses options which are less harmful to the environment while he produces or consumes in his daily life. Environmental education is necessary in order to make use of the environment in a reasonable way to meet some needs, and to prevent consumption and pollution caused by inappropriate use of the environment [Şafak, Erkal,1999].

Today, one of the most frequently discussed issues regarding the control of the environment, is the dilemma caused bye differentiation and even alienation between economical development and protection of the environment [Şafak et al.,2008].

Sustainable development is defined as "Humanity has the ability to make development sustainable to ensure that it meets the needs of the present without compromising the ability of future generations to meet their own needs" [Kates et al.,2005].

The sustainable development concept was first mentioned in the report prepared by the ex Prime minister of Norway, Groharlem Brundtland, who was the chairperson of the committee commissioned by the UN Environment Program for the preparation of this report for the Third of 10th World Years of Environment. The purpose of this report was to find a solution for the dilemma between development and the protection of the environment, which is being discussed mostly in the public opinion for environmental issues. It is stated that the development and the protection of environment concepts do not exclude each other, to the contrary, but they support each other and it is possible to carry on with the development efforts which the countries need and it is possible to do it without damaging their environment. The basic principle of sustainable development is to guarantee to keep the use of natural resources within a limit not exceeding their natural reproduction speed [Kahraman & Türkay,2006;Yıldız, Sipahioğlu & Yılmaz, 2005, TÇSV,1991].

Environmental Education

Environment education develops skills, knowledge and values that promote sustainable environmental management [Şafak et al.,2008].

Today to cope with the changing environmental conditions and the environmental problems, countries separate their budgets in to the solving these problems. But; the easiest way is that growing up sensible generation about environmental problems. Especially environmental education has an important role in order to realize this mission. Environmental education as aformal education movement has its origins in the concerns about environmental degradation and decreasing quality of life expressed by scientists in the 1960s. Through the decade of the 1970s and into the 1980s, the environmental education movement grew rapidly. Education concerning environmental problems was recognized at the World Conference on the Environment in Rio de Janeiro in 1972. These emergences are today evident in environmental education [Şafak, Erkal,1999; Tuncer et al.,2005]. Education plays an important role in environmental awareness at different levels [Kiziroğlu, 2007].

It is well known that education and training activities are important to find long-lasting solutions to environmental problems. The most effective method to be adopted to solve these problems is to raise awareness of environmental issues [Uzun and Sağlam 2006, 240]. The main focus of environmental education programs has been to change environmental behavior through increasing environmentalknowledge [Pooley and O'Connor,2000].

Environmental education is a new way of learning about human relations with the environment. The object of environmental education is focused on the relationship and impact of humanity on the environment. It is the type of education stressing an integrated way of structuring human nature and natural and physical resourses (Tuncer and Erol, 1992). The aim of environmental ducation is to literate students for playing active erole in preventing environmental problems and be aware or sensitive about environmental issues [Özden,2008].

Human being, desiring to put their tomorrow under guaranty, started to be more sensitive to the environmental problems and these problems appeared in the agenda of the societies with different events and thus uncovered the consciousness towards environment. The interest shown to the environment, in view of the limited natural resources, reflects the importance of the sensitive balance between the human being and the environment. The importance of environmental awareness, environmental sensitiviness and environmental consciousness is indisputable for healthy human-nature relationships. The social roots of the coming global ecological crisis lie largely in the very structures of the present awareness of society for environment and values inherent in the way they function (Cankurt et al.,2008). It is a necessity to inform individuals about environmental issues and raise ecological awareness among them in such way to ensure behavioral changes The purpose of environmental education and training is to provide individuals with the ways and means to effectively, responsibly protect the environment [Çelen et al. 2008]

The role of education in conservation and in addressing the rapidly increasing environmental problems, while improving the environment, is well-known today.

Environmental education needs a generally accepted definition, since there are different definitions and interpretions. A systematic approach to education on environmental needs must contain the following:

- Environmental education must be comprehensive,
- It must include the inter-relations of parts and elements of natural life and global structure, and
- The problems and causes of pollution must be approached from the point of view of the aforementioned relations [Tuncer and Erol, 1992].

Pricipies of Environmental Education:

- Environmental educationmust involve everyone
- Environmental education must be lifelong
- Environmental education must be holistic and about connections.
- Environmental education must be practical
- Environmental education must be in harmony with social and economic goals and accorded equal priority [EEU,2000].

The Turkish education system is subject to severe alterations at present. Among others, environmental education andd sustainable development principles are introduced at all levels. An elective cource "Nature and Mankind" is taught in the ninth class. Environment related topics are not only presented theory, but the pupils have also the opportunity to learn them in practice outside the classroom. According to the framework plan for primary schools, environmental education is integrated into social studies., biology, chemistry, and sciences etc., and no longer taught only within the subject "Environment, Health, Traffic, and reading" as it was the case previously. The following goals shall be anchored in the consciousness of the students:

- Learning the social, biological, and physical factors of environment.
- Respect for Nature and Mankind
- Recognition of environmental degradation for human beings and the society
- Development of a consciousness for environmental protection,
- Drawing attention on the impact of environmental problems on human health[Kiziroğlu, 2007].

Environmental subjects in the central cources are structured in a manner that, in dealing with the environment, cources on life, social and scientific knowledge were planned to cover environmental problems. According to the level of classes, subjects would be selected from the following:

- Protection of soil, water, air,
- Solutions to pollution,
- Ways of giving guarantees for living without risks,
- Conservation of natural resources [Tuncer and Erol, 1992].

Environmental education and awareness are implemented in the public and the private school systems in Turkey [Tuncer et al., 2005].

The number of Universities in Turkey is more than 100, distributed throughout the country. They are autonomous while setting new study cources and deciding on research topics in environmental sciences. Some Universities have also started environment cources and founded research centers for environmental sciences in order to work on environmental problems and to act as advisory boards for public authorities in Turkey [Kiziroğlu, 2007].

Environmental Attitudes of Young People and Some Studies

The ecological attitudes adopted today will shape the environment of the future [Özdemir et al. 2004]. Education and training of particularly young people will guarantee conservation for the future of societies [Öztek 2006]. "Environmental attitudes of young people" is a very important category because the youth will be more effective in finding solutions to the environmental problems produced by current conditions. Generally, individuals start to develop ecological attitudes at very early ages. Many children develop their own ideas about some environmental issues such as energy and technology when they reach adolescence. Increasing students' knowledge about environmental issues ensures positive improvements in their ecological attitudes. Some researchers suggest that student participation in ecology classes produces an increase in their responsible behavior and an increase in their awareness about environmental problems while some others suggest that development of environmental attitudes is based on life experience rather than participation in some specific academic courses [Bradley et al. 1999, 17].

Studies show that positive ecological attitudes are dependent on the age factor, and young people know more than old people about ecological issues [Ewert and Galloway 2004; Koval, Merting 2001]. Eagles and Demare (1999) showed that there was no relationship between gender and environmental sensitivity and awareness, but their results showed that females have a more moralistic attitude toward the environment than males do. Berberoğlu and Tosunoğlu (1995) found that population growth, nuclear energy conservation and environmental problems were seen as the important environmental issues for university students in Turkey.

There is substantial research on environmental attitudes: Gökçe et al. [2007]; Uzun and Sağlam [2006); Özmen, Çetinkaya and Nehir [2005]; Yılmaz et al. [2002]; Bradley, Waliczek and Zajicek [2001]; Pooley and O'Connor [2000]; Geok et al. [1998]; Tikka et al[2000]; Chan [1996]; Berberoğlu and Tosunoğlu [1995]; Tosunoğlu [1993]; Schmidt, [2007]. However, there are only a limited number of studies on the environmental attitudes of university students.

Therefore, this chapter was designed and conducted to determine university students' attitudes about the environment and environmental problems.

METHOD

Sample

No sample section was made in the scope of this descriptive study -which was conducted on students enrolled in the Department of Family and Consumer Sciences, Faculty of Industrial Arts Education, Gazi University (n: 358); rather, respondents who agreed to complete the questionnaire were included in the study's scope (n: 300).

Instruments

Personal Data Questionnaire

In the first part of the study, we administered the questionnaire. This part includes students' personal data such as grade, age, sex, number of siblings; any ecological education received before university; any out-of-school environmental training; their opinions about the proper way to assess their understanding of the environmental studies course, the responsibility of individuals, state and conservationist institutions in the solution of environmental problems, and the world's most important environmental problems.

Environmental Attitude Scale

In the second part of the study, a 5-point Likert-Type "*Environmental Attitude Scale*" developed by Berberoğlu and Tosunoğlu (1995) was used to determine students' opinions about environmental problems. This 21-item scale had 0.77 Cronbach Alfa Coefficient reliability. In this study, the scale's Alfa internal reliability was found to be 0.84. It is in the mid-level reliability range. The lowest possible score to be taken from the scale was 21 and the highest was 105. Environmental awareness increases in proportion to the increase in the scores taken from the scale.

The mean of items varied between 4.05-1.31 and their standard deviation between 1.38-0.66. According to these results, the highest mean attitude was recorded in Item 10 (Air, water and soil are inexhaustible resources) while the lowest mean attitude was recorded in Item 3 (Thinning ozone layer threatens all humanity). Upon examination of the text-item correlation showing the validity values of items, item validity values were found to vary between 0.62 -0.23. In light of these results, each item was at an acceptable validity level (figure 5).

Arrangements on the basis of data:

1. Since there were only 5 individuals in the "26+" age group, this group was combined with the "23-25" group. Thus there were only two groups, which enabled the T-test.
2. Since there were 14 individuals in the "2251+" Turkish Liras income group, this group was combined with the "1501-2250" liras income group. Thus, a new group was categorized as "1501+" and variance analysis was made.
3. The variable of "the number of siblings" was reduced to two groups divided by those having less than three siblings and those having more than three siblings.
4. Students who gave an answer of "no idea" or "unnecessary" to "necessity of environmental education in primary, secondary or tertiary education" were combined and T-test was conducted.

Data Analysis

Study data were collected by using the questionnaire developed by researchers. The questionnaire was distributed to 358 students between 30 September 2008 and 08 October 2008, and then the completed forms of 300 respondents were collected. The SPSS 13.0 statistical program was used to analyze the data. Mean, standard deviation, T-test, Levene test, Tamhane test and LSD test were used in the analysis. The groups with meaningful Levene test results were compared with Tamhane test while the groups with no meaningful Levene test were compared with LSD test.

Study Limitations

Main limitation was that although it was aimed to access all students in the study universe rather than making a sample selection, it was impossible for various reasons (incomplete forms, student hesitation to fill in the forms, etc). The study's participation rate was 83.8%. Moreover, the study was limited to students registered in the Department of Family and Consumer Sciences, Faculty of Industrial Arts Education, Gazi University.

CONCLUSION

Examination of figure 1 shows that the respondents' scores varied in the 21-87 range, with a mean value of 55.09. Taking into consideration the calculated attitude mean and that the maximum score to be taken from the test was 105, we can conclude that respondents showed a mid-level attitude. Low standard error value provides information about the reliability of the results. Furthermore, considering the scale's skewness and kurtosis value, the distribution was found to be under left skew, most of the scores were above the mean value and there was a partial kurtosis. Explanations can also be examined visually in the total score histogram (figure 2).

N	300.0
Mean	55.09
Std. Error	0.69
Median	59.00
Mod	61.00
Std. Deviation	12.02
Skew ness	-0.97
Kurtosis	0.46
Minimum	21.0
Maximum	87.0

Figure 1. Descriptive statistics about total attitude scale scores.

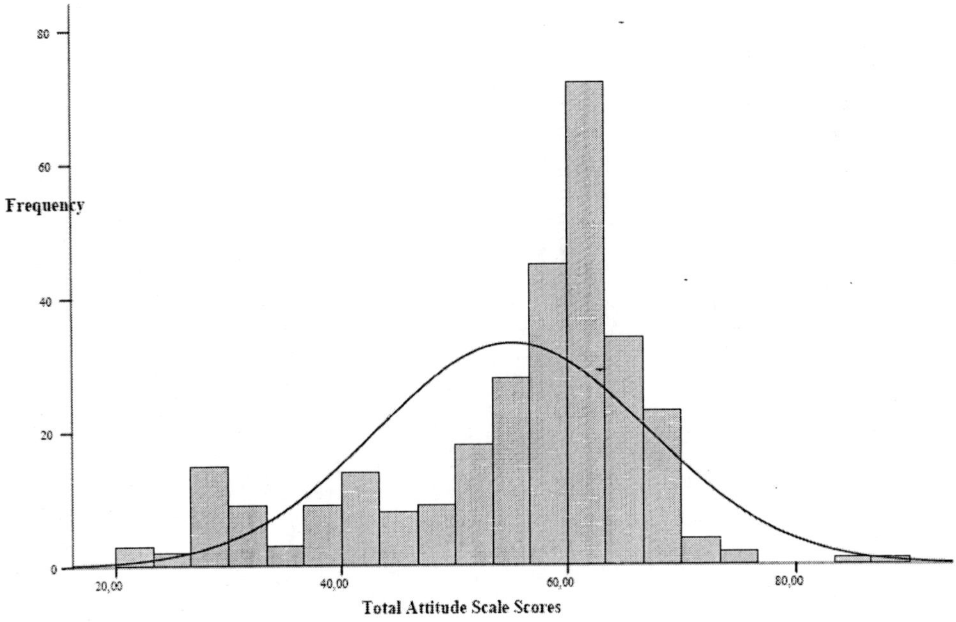

Figure 2. Graphic of total attitude scale scores.

Variable		Group	N	Mean	Std. Deviation	T
Age		18-22	210	55.22381	11.79216	
		23-26	90	54.77778	12.60085	0.29
Sex		Male	137	55.87	12.20	
		Female	163	54.44	11.87	1.03
No. of siblings		Less then three	142	55.82	12.48	
		Three or more	158	54.44	11.59	0.99
Any ecological education before university		Yes	116	55.72	11.65	
		No	184	54.09	12.24	-1.15
Any out-of-school ecological training		Yes	32	52.59	12.47	
		No	268	55.39	11.96	-1.24
Opinions about the right grade to give environmental course	Primary Education	Necessary	268	56.18	11.21	
		Not Necessary	32	45.97	14.66	4,7**
	Secondary Education	Necessary	269	56.19	11.19	
		Not Necessary	31	45.55	14.70	4,84**
	Tertiary Education	Necessary	267	56.36	11.06	
		Not Necessary	33	44.85	14.53	5,43**
Responsibility of individuals for the solution to environmental problems		Too much	215	55.15	11.68	
		Much	85	54.93	12.90	0.15
Responsibility of the state for the solution to environmental problems		Too much	220	55.37	11.92	
		Much	80	54.33	12.33	0.66
Responsibility of conservationist institutions for the solution to environmental problems		Too much	209	54.97	11.80	
		Much	91	55.36	12.57	-0.26

** $P<0,05$.

Figure 3. Independent groups T-test results of two-category variables.

Examination of figure 3 reveals no statistically meaningful difference between the mean attitude scores of the groups relevant to the variables of age, sex, number of siblings, any environmental education received before university, any out-of-school environmental training and the responsibility of individuals, state and conservationist institutions for the solution to environmental problems. These variables are parallel to the environmental attitude means of students. This chapter revealed that the groups' mean attitude values were similar when controlling the variables of age, sex, number of siblings, any environmental education studied before university, any out-of-school environmental training and the responsibility of individuals, state and conservationist institutions for solving environmental problems. These results suggest that university students of different age and sex have similar attitudes towards this issue. Studies conducted by Özmen et al. [2005], Çelen et al. [2008], Şama [2003], Hampel et al. [1994], on the other hand, suggest that females are more sensitive to environmental problems than males.

However, the groups' mean values significantly differed when considering the variable of the appropriate assessment for students in the environment course. In the primary school variable, the mean score of students who think that the environment course should be given in primary school was found to be higher than that of respondents who believe there is no need for an environment course in primary school. In the secondary education variable, the mean score of the students thinking that an environment course should be studied in secondary school was found to be higher than the mean of those who feel no need for such a course in secondary school. In the tertiary education variable, we found that the mean score of respondents who think that an environment course should be studied in universitiesis higher than that of students who consider no need for a class in environment course at universities.

The mean score of the respondents who think that an academic course in environment should be studied in primary, secondary or tertiary education was higher than that of students stating that it is unnecessary to study an environment course in primary, secondary or tertiary education. This result shows that students want ecological studies included as part of the curriculum at these three educational levels. No statistically meaningful difference was found between the mean attitude scores of students who received environmental education before university and those who did not. The chapter conducted by Çelen et al. [2008] also revealed no difference between the environmental awareness scores of students who took environment-related courses in primary school, secondary school and high school and those who did not.

Examining figure 4, we can conclude that there is a meaningful difference between the environmental attitude scores of groups in terms of academic year and income variables. The mean score of students in the 0–750 YTL income group was found to be higher than that of students in the 751–1500 and 1501-higher income groups. The mean environmental attitude score of the 1st year university students was recorded higher than that of 4th year university students. This finding can result from the fact that most of the 1st year university students took an environmental courses in their pre-university years. Şama's study [2003], on the other hand, showed that being in the first or the fourth year has no effect on students' environmental attitudes.

Variable	Group	N	Mean	Std. Deviation	Std. Error	Levene	F	Difference
Academic Year	1	76	57.72	8.55	0.98	8.24**	2.73*	1 to 4
	2	71	55.83	11.46	1.36			
	3	73	54.58	12.63	1.48			
	4	80	52.40	14.18	1.59			
The Place Where They Live	Center of a province	161	55.91	11.82	0.93	0.7	0.82	NA
	Center of a district	102	54.10	12.70	1.26			
	Village/Town	37	54.24	10.95	1.80			
Income	1. 750	77	58.70	6.34	0.72	23.65**	6.01**	1 to 2 1 to 3
	2. 751-1500	137	54.80	12.63	1.08			
	3. 1501 +	86	52.31	14.05	1.51			
The Most Important Environmental Problem in the World	1.Air, water pollution	126	55.62	11.85	1.06	2.63	1.84	NA
	2-Population increase, urbanization	118	53.58	12.71	1.17			
	3.Deforestation	56	57.09	10.65	1.42			

*p < 0,05; ** p < 0,01.

Figure 4. Variance analysis results for more-than-two group variables.

Items	Mean Value	Std. Deviation	Test-item correlation
10- Air, water and soil are inexhaustible resources.	4.05	1.11	0.39
7- Efforts exerted on Turkey's southern coasts to save the turtles are waste of time.	3.79	1.08	0.28
1-It is unnecessary for the World Bank to support air pollution measurement projects in Turkey because there are much more important projects to be supported in Turkey.	3.74	1.26	0.53
2-Use of natural gas in houses and workplaces will not solve the air pollution problem.	3.65	1.24	0.55
8-Turkey does not have any desertification problem.	3.59	1.38	0.62
16-Environment conservation is an idea created only to prevent the development.	3.59	1.29	0.55
15-Slum is not an environmental problem.	3.50	1.35	0.50
5-News about pollution of seas, rivers and lakes is exaggerated.	3.46	1.27	0.42
18- Environmentalists form groups to meet the need for new friends rather than protecting the environment.	3.08	1.31	0.48
9-Construction of small houses in the forested lands close to cities, so as to satisfy people's fresh air need, should be supported.	2.88	1.43	0.59
19-No institution should interfere in a country's sovereign use of natural resources.	2.46	1.47	0.39
6-Drinking water used in big cities is so polluted that people have to use filters in their houses.	2.01	1.31	0.31
13-Malnutrition in underdeveloped countries is a result of environmental problems.	1.95	1.05	0.39
4-Meetings should be organized to protest technological products that damage the ozone layer.	1.89	1.05	0.42
17-Environmental awareness does not prevent a country's development.	1.89	1.00	0.39
11-Any country making nuclear trials should be protested.	1.75	1.06	0.27
12-Rapid population growth is an important environmental problem.	1.73	0.98	0.33
14-People who litter and spit should be reprimanded.	1.71	1.13	0.32
21-Environmental Ministries should be established to solve countries' environmental problems.	1.60	0.88	0.30
20-Environmental problems should be given more space in newspapers, periodicals and television.	1.48	0.77	0.28
3-Thinning ozone layer threatens all humanity.	1.31	0.66	0.23

Figure 5. Mean values, standard deviation and test-item correlations for attitude scale items.

To determine the groups between which there were differences, the groups' homogeneity was the first factor examined. Groups with meaningful Levene test results were compared on the basis of Tamhane test while the groups with non-meaningful Levene test results were compared according to LSD test. According to Tamhane test results conducted for the academic year variable, the mean environmental attitude score of 1st year university students was higher than that of 4th year university students.

According to Tamhane test results conducted for income variable, the mean score of the students in the 0–750 Turkish Liras (YTL) income group was higher than that of students in the 751–1500 and 1501-more income groups.

This chapter shows that there was no statistically meaningful difference between mean "Environmental Attitude Scale" scores of students living in the center of a province. Şama's study, on the other hand, found that students living in larger settlement areas developed environmental attitudes stronger than those living in smaller settlement areas.

Respondents listed air and water pollution (42.0%), population growth and urbanization (39.3%) and deforestation (18.6%) as the three most important environmental problems. The fact that students listed air and water pollution, rapid population growth, urbanization and deforestation as the world's most important environmental problems is an indicator that students are aware of environmental problems. The study conducted by Kasma-Ronkainen and Virokannas [1996] on university students in Finland showed that more than half of the students listed water pollution and deforestation as the two most important environmental problems. The report prepared by Çelen et al. [2002] revealed that students ranked air pollution, environmental pollution and wastes as the most important environmental problems.

The fact that "air, water and soil are inexhaustible resources" item produced the highest and "thinning ozone layer threatens all humanity" item produced the lowest mean environmental attitude scores, and that "people who litter and spit should be reprimanded" item and "ecology-related news should be given more space in newspapers, periodicals and television" item were given low environmental attitude scores can lead us to believe that students do not attach importance to environmental problems and are not informed enough about environmental issues.

As a conclusion; "Environmental Attitude Scale" scores of respondents varied between 21-87 points. No statistically meaningful difference was found between the mean attitude scores of groups when controlling variables of age, sex, number of siblings, any environmental education received before university, any out-of-school environmental training and the responsibility of individuals, state and conservationist institutions for solving environmental problems. However, opinions among the groups significantly differed when considering the variable of how to assess students in the environmental education course. The mean environmental attitude score of the first year students was recorded to be higher than that of the seniors. The mean score of students in the 0–750 YTL income group was higher than that of respondents in the 751–1500 and 1501-more income groups. Considering the respondents' "Environmental Attitude Scale" scores, the highest mean attitude was recorded in Item 10 (air, water and soil are inexhaustible resources) while the lowest mean attitude was recorded in Item 3 (thinning ozone layer threatens all humanity).

The following recommendations are made based on these results:

- Environmental education should be provided in all formal and informal education programs; thus, students should be informed about the reasons for and ways to prevent environmental problems;
- An environmental education curriculum should be prepared in such way as to positively influence environmental attitudes and behavior;
- More research should plan about environmental issues.

REFERENCES

Berberoğlu, G. & Tosunoğlu C. (1995). Exploratory and Confirmatory Factor Analyses of an Environmental Attitude Scale (EAS) for Turkish University Students. *Journal of Environmental Education. 26 (3)*, 40-44.

Bradley, J.C., Waliczek, T.M.& Zajicek, J. M. (1999). Relationship between Environmental Knowledge and Environmental Attitude of High School Students. *Journal of Environmental Education. 30 (3),*17-21.

Cankurt, M et al.(2008). Awareness to Environmental Pollution in Turkey. Southern Agricultural Economics Association Annual meeting, Dallas, February,2-6, 2008.

Chan K. (1996) Environmental Attitudes and Behaviour of Secondary School Students in Hong Kong. *The Environmentalist. 16 (4)*, 297-306.

Çelen, Ü et al. (2008). Ankara Üniversitesi Sağlık Eğitim Fakültesi Öğrencilerinin Çevre Duyarlılığı ve İlişkili Faktörler. Available from: [http://www.dicle.edu.tr/halks/m69.htm] (Available date:22.10.2008)

Dastan, H. (1999). Çevre Koruma Bilinci ve Duyarlılığının Oluşmasında Eğitimin Yeri ve Önemi (Türkiye Örnegi), Gazi Üniversitesi, Sosyal Bilimler Enstitüsü, Ankara,pp 14.

Eagles, P.F.J and Demare, R. (1999). Factors influencing children's environmental attitudes. The Journal of Environmental Education. 30 (4),33-37.

EEU. (2000).Environmental Education for a Sustainable. Environmet Education Unit Australia.Department of the Environment and Heritage. Available from: Future.http://www.environment.gov.au/education/publications/nap/pubs/nap.pdf

Erol, G.H. & Gezer, K (2006). Prospective of Elemantary School Teachers' Attitudes Towar Environment and Environmental Problems. International Journal of Environmental and Science Education. 1(1),65-77.

Ertürk, H. (1996). Çevre Bilimlerine Giriş, Uludağ Üniversitesi Güçlendirme Vakfı Yayınları No: 10, Bursa, 1996,pp 9.

Ewert, A. & Galloway, G. (2004). Expressed Environmental Attitudes and Actual Behavior: Exploring the Concept of Environmentally Desirable Responses. Available from:[http://www.latrobe.edu.au/oent/OE_conference_2004/papers/ewert2.pdf] (Available date: 24.10.2008).

Geok, T. et al. (1998). A Survey of Environmental Knowledge, attitudes and Behaviour of students in Singapore. *International Research in Geographical and Environmental Education. 7(3)*, 181-202.

Gökçe, N et al. (2007). Elementary Students Attitudes Towards Environment. *Elementary Education Online. 6(3)*, 452-468

Hampel, B., Holdsworth, R. & Boldert, J. (1994). Urban/Rural Diffirences in Environmental Consciousness Among Adolescents. Available from: [http://www.csu.edu.av /research /crs / ruralsoc/ v5n4p13 . htm] (Available date:18.11.2008)

Kabaş, D. (2004). Kadınların Çevre Sorunlarına İlişkin Bilgi Düzeyleri ve Çevre Eğitimi. Yayınlanmamış Yüksek Lisans Tezi . Gazi Üniversitesi , Eğitim Bilimleri Enstitüsü , Aile Ekonomisi ve Beslenme Anabilim Dalı.

Kahraman, N.,& Türkay, O. (2006). *Turizm ve Çevre*. Detay Yayıncılık, Ankara.

Kasma- Ronkainen, L. & Virokannas H. (1996). Concern About the Environment Among Medical Students. *Scand J. Soc Med. 24,* 121-123.

Kates,R.W., Parris,T.& Leiserowitz, A.A (2005). What is Sustainable Development? Goals, Indicators, Values, and Practice. Environment: Science and Policy for Sustainable development,47(3),8-21.

Kiziroğlu, I. (2007). Education and Research on Environmental Awareness in Turkey. Clean. 35(6),534-536.

Koval, M.H & Mering A.G (2001). Attitudes Toward Natural Resources and Their Management : A Report on the "2001 Resource Attitudes in Michigan Survey" Available from:[http:www.michigandnr.com/publications/pdfs/huntingwildlifehabitat/WCS/1_7_cit ions.pdf], (Available date:18.11.2008).

Özdemir, O, Yıldız, A & Ocaktan, E & Sarışen, Ö (2004). Tıp Fakültesi Öğrencilerinin Çevre Sorunları Konusundaki Farkındalık Ve Duyarlılıkları. *Ankara Üniversitesi Tıp Fakültesi Mecmuası. 57(3),*117-127.

Özden, M (2008). Environmental Awareness and Attitudes of Students teachers: An Empirical Research. International Research in Geographical and Environmental Education. 17 (1).40-55.

Özmen, D., Çetinkaya, A.Ç. & Nehir, S. (2005). Üniversite Öğrencilerinin Çevre Sorunlarına Yönelik Tutumları. *TSK Koruyucu Hekimlik Bülteni. 4 (6)*, 330-344.

Öztek, Z. (2006). İlk ve Ortaöğretimde Çevre Eğitimi. II. Çevre hekimliği kongre kitabı. 18-21 Ocak, Ankara.210-212.

Pooley, J.A. & O'Connor, M. (2000). Environmental Education and Attitudes, *Environment & Behavior. 32 (5)*, 711-724.

Schmidt, J, E.(2007). From Intentions to Actions: The Role of Environmental Awareness on College Students. Journal of Undergraduate Research. Available from: http://www.uwlax.edu/URC/JUR-online/PDF/2007/schmidt.pdf.

Soran, H. et al. (2000). Biyoloji Öğrencilerinin Çevre Konularına Olan İlgilerinin Araştırılması ve Kimya Öğrencileri ile Karşılaştırılması. *Hacettepe Üniversitesi Eğitim Fakültesi Dergisi. 18*, 128-139.

Şafak, Ş.,& Erkal, S. (1999).Çevre Eğitimi ve Aile. *Eğitim ve Bilim Dergisi.* 23(112),63-66.

Şafak, Ş, Yertutan, C& Erkal,S (2008). Sürdürülebilir Kalkınmada Çevre Eğitimi ve Aile, *Blacksea International Environmental* Symposium, 25-29 August 2008, Giresun–Turkey.

Şama, E. (2003). Öğretmen Adaylarının Çevre Sorunlarına Yönelik Tutumları. *G.Ü. Gazi Eğitim Fakültesi Dergisi.* 23(2), 22-25

TÇVS (1991). Ortak Geleceğimiz. Dünya Çevre ve Kalkınma Komisyonu, Türkiye Çevre ve Kalkınma Komisyonu, Üçüncü Baskı, Önder Matbaa.

Tikka, P. M., Kuitune, M.T & Tynys, S.M (2000). Effects of Educational Background on Students' Activity Levels, and Knowledge Concerning the Environment (Statistical Data Included), Journal of Environmental Education, 31, 12-20.

Tosunoğlu G. (1993). A Study on the Dimensions and Determinants of Environmental Attitudes. Middle East Technical University. PhD thesis.

Tuncer, M & Erol, D (1992). The Environmental Education in Turkey: Some View and Proposals of Biopolitics. 5th *BIO International Conference* 7-10 May, Istanbul Available from: http://www.biopolitics.gr/HTML/PUBS/VOL4/fe-tunce.htm.

Tuncer, G, Ertepinar H, Tekkaya C& Sungur S. (2005). Environmental Attitutes of Young People in Turkey: Effects of School Type and Gender. Environmental Education Research, 11(2), 215-233.

Uzun, N.& Sağlam, S. (2006). Ortaöğretim Öğrencileri İçin Çevresel Tutum Ölçeği Geliştirme ve Geçerliliği. *H.Ü. Eğitim Fakültesi Dergisi. 30(2006),* 240-250

WHO, (2008). Global Environmental Change. Available from: http://www.who.int/globalchange/en/

Yıldız, K, Sipahioğlu, Ş & Yılmaz, M (2005). Çevre Bilimi. Gündüz Eğitim ve Yayıncılık. Ankara. pp 9.

Yılmaz, A et al. (2002). Ortaöğretim ve Üniversite Öğrencilerinin Çevre Kavramları ve Sorunları Konusundaki Bilgileri ve Öneriler. *Hacettepe Üniversitesi Eğitim Fakültesi Dergisi. 22,* 156-162.

Yücel, F. (2004). Sürdürülebilir Kalkınmanın Sağlanmasında Çevre Korumanın ve Ekonomik Kalkınmanın Karşıtlığı ve Birlikteliği. Çukurova Üniversitesi, Sosyal Bilimler Enstitüsü Dergisi. Sayı:11.

In: Environmentalism: Environmental Strategies... ISBN: 978-1-60741-634-0
Editors: E. Kahraman and A. Baig, pp.169-182

Chapter 8

INFLUENCE OF FOREST DISTURBANCE ON WATER DISCHARGE

Koji Tamai
Forestry & Forest Products Research Institute, Tsukuba, Japan

ABSTRACT

We examined the influence of two causes of forest decline, i.e., forest fire and pine wilt disease, on discharge duration curves in a paired-catchment experiment in the Tatsunokuchi-yama Experimental Forest, Japan (34°42'N, 133°58'E). Regression lines between discharge duration curves for the Kita-tani and Minami-tani catchments were calculated during the control period (36 years). The observed discharge was then compared to that estimated from the regression lines for the disturbed period (19 years). Disturbance generally increased daily discharge, except on several dozen days. The observed discharge was increased by approximately 15%, 35%, and 43% for days 1–40, 140–320, and more than 320, respectively, in the discharge duration curve.

Further, from a comparison of the discharge duration curves of the two catchments, we developed a new index for estimating the short-term influence of small forest disturbances on discharge rates in the Minami-tani catchment. We used this index to estimate the influence of disturbance, and the recovery speed from that disturbance, in a 22.61-ha catchment. The results showed the following:

1. Brush cutting increased the discharge rate when the tree volume was very small.
2. Establishment of a hinoki (*Chamaecyparis obtusa*) plantation, combined with a lack of live stem clearing, and hinoki thinning, did not influence the discharge rate.
3. The influence of pine die-off on low and drought water discharge remained even 20 yr after disturbance, probably because of the absence of afforestation following the disturbance.
4. Afforestation following disturbance led to a rapid decrease in the discharge rate.

1. INTRODUCTION

Forests influence water discharge via transpiration and infiltration into deeper soil. The former process results in decreased discharge rates, especially peak discharge, and thereby contributes to flood prevention. The latter process delays water discharge and is thought to increase base flow. Therefore, forests affect base flow antithetically, by decreasing base flow through transpiration and increasing it by infiltration. The relationship between forest characteristics and base flow rate varies with geological, topological, and meteorological features.

Many previous studies (e.g., Fahey et al., 1997; Fujieda & Abe, 1982) have simply categorized catchment vegetation as "forest" and "extinct forest," even though forests actually develop year by year and likely have varying influences on discharge according to their successional stage. Moreover, the influence of relatively small disturbances should also be estimated. In this study, we propose a new index for estimating the relationship between forests and annual discharge. This index allows us to estimate how discharge is variously influenced by forest recovery from differing degrees of disturbance.

2. SITE DESCRIPTION

2.1. Location

The Tatsunokuchi-yama Experimental Forest is located at 34°42' N, 133°58' E in Okayama Prefecture, Japan. The experimental forest consists of two adjacent catchments, Kita-tani (KT, 17.27 ha) and Minami-tani (MN, 22.61 ha; figure 1), and is mainly underlain by Paleozoic formations. Discharge and precipitation have been measured at or near the site since 1937; discharge is measured in the forest using a wire gauge, and precipitation is measured at an adjacent weather station using a tipping bucket. Annual precipitation and mean air temperature are approximately 1200 mm and 14.3°C, respectively, and there is no seasonal snow cover

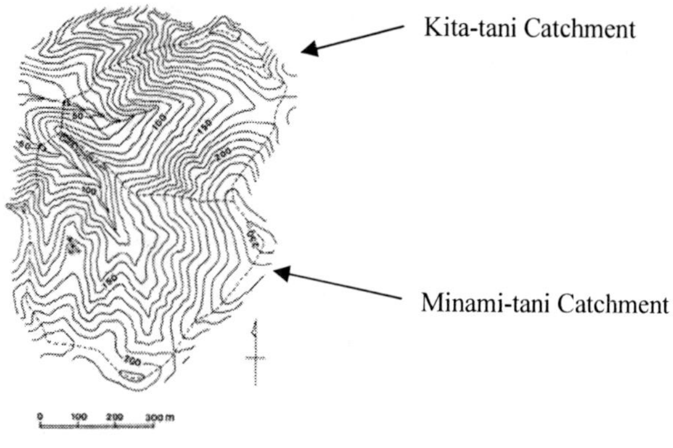

Figure 1. Topography of Kita-tani and Minami-tani catchments in Tatsunokuchi Experimental Forest.

2.2. Vegetation History

Red pine (*Pinus densiflora*) was the dominant species at both catchments in 1937 when observations first began. The forest began declining in 1940 as a result of pine wilt disease, and timber was harvested in all areas of both the KT and MN catchments. Natural regrowth of red pine, oak (*Quercus serrata*), and bamboo (*Pleioblastus chino*) occurred in both catchments, which had the same vegetation cover until 1954. Since that time, deciduous broad-leaved trees have continuously dominated KT, but vegetation in MN has been disturbed many times (table 1). First, in November 1954, brush in a 7.5-ha area of MN was cut and removed, and hinoki (*Chamaecyparis obtusa*) was planted at a density of 3000 stems ha^{-1}. In September 1959, a forest fire burned the forest and litter in 22.3 ha of MN, and pine was planted in March 1960 at 3000 stems ha^{-1}. MN experienced another forest fire in 1974 that burned 3.45 ha, and 0.35 ha of forest were harvested in 1977. Hinoki was planted after both cases. Until 1978, natural thinning reduced the pines in MN to 1250–2650 stems ha^{-1}. From 1978–1980, around 96.5% of the remaining pines, mainly in MN, died from pine wilt disease (Kobayashi et al., 1981), and the forest in MN nearly disappeared, except for the area planted with hinoki. No afforestation occurred following the removal of the dead pines, and natural regrowth of broad-leaved trees occurred. In 1982, a 2.50-ha area was planted with hinoki in MN after removal of the dead pines. Subsequently, 2.34 ha (1997) and 2.53 ha (2000) of hinoki forest in MN were thinned artificially.

Table 1. Disturbance history in Tatsunokuchi-yama Experimental Forest

Year	Catchment	Disturbance Type	Area (ha)	Treatment
1944-47	KT	Clear cut	17.27	Natural Regrowth
1944-45	MN	Clear cut	22.61	Natural Regrowth
1954	MN	Brush cutting	7.5	Hinoki Planted in 1955
1955	MN	Brush cutting	7.2	Hinoki Planted in 1956
1957	MN	Brush cutting	4.8	Hinoki & Pine Planted
1959	MN	Forest fire	22.3	Pine planted in 1960
1962	KT	Riparian clear cut	0.4	Natural Regrowth
1964	KT	Riparian clear cut	0.4	Natural Regrowth
1974	MN	Forest fire	3.45	Hinoki Planted in 1976
1977	MN	Clear cut	0.35	Hinoki Planted
1978-80	MN	Pine Wilt Disease	18.8	Natural Regrowth
1982	MN	Brush cutting	2.5	Hinoki Planted
1998	MN	Hinoki Thining	2.34	
2001	MN	Hinoki Thining	2.53	

3. Estimating the Effect of Forest Decline in the Minami-Tani Catchment

3.1. Method

3.1.1. Paired-Catchment Experiment

A paired-catchment experiment can be used to estimate the effect of forest conditions on shares of water discharge and evapotranspiration. In this method, discharge volumes are compared from adjacent catchments during control and disturbed periods. In the control period, both catchments should have almost the same forest conditions. In the disturbed period, one of the catchments should have experienced forest disturbance. The relative discharge change can then be estimated for the disturbed catchment by comparison to discharge in the control catchment during a certain period. The discharge volume and duration will vary depending on the soil moisture conditions just before precipitation events. Precipitation patterns affect discharge volume and its fluctuations. A single-catchment experiment can show the absolute discharge change between before and after forest disturbance. However, this method cannot reveal the effect of forest disturbance and precipitation patterns and provides vaguer conclusions (Bosch & Hewlett, 1982).

Abe and Tani (1985) and Hattori (1994) conducted paired-catchment experiments in the Tatsunokuchi-yama experimental forest. They compared paired annual discharge volumes and estimated the effects of forest fire and pine wilt disease. The difference between the annual discharge from the two catchments (ΔQ) was positive in almost all years, as calculated by Eq. (1) (Hattori, 1994):

$$\Delta Q \quad Q_{KT} \quad Q_{MN} \tag{1}$$

where Q_* is the annual discharge volume and KT and MN indicate the Kita-tani and Minami-tani catchments, respectively.

However, ΔQ was negative in 1960–1966 and 1981–1994 (figure 2). The negative value can be explained by decreased evapotranspiration in the MN catchment due to a forest fire in 1959 and pine wilt disease from 1978–1980.

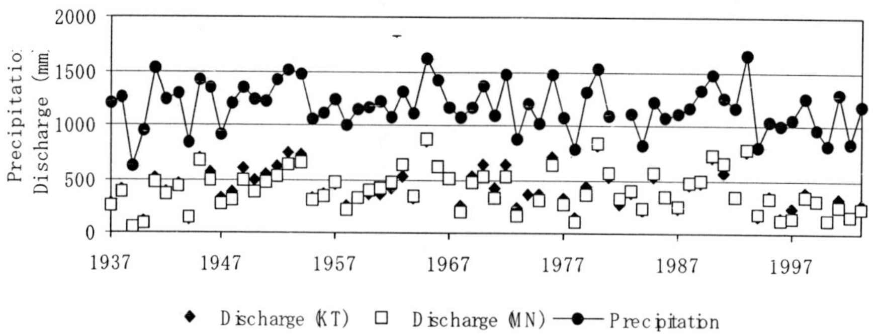

Figure 2. (Continues)

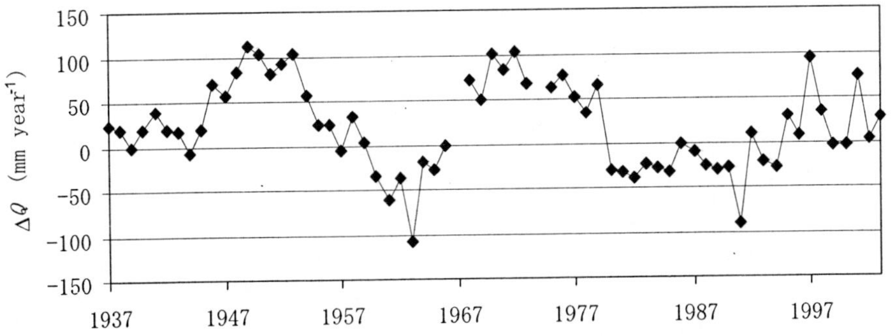

Figure 2. Annual discharge and precipitation in Tatsunokuchi-yama Experimental Forest.

In this study, we defined the disturbed periods as 1960–1966 and 1981–1994 when ΔQ was negative, the damage periods as 1959 and 1978–1980 when disturbances occurred, and the control period as all other years in the period 1937–2003, respectively. The MN catchment represented the disturbed catchment, while the KT catchment served as the control. Data from years with missing observations were excluded. Thus, there were 19 years for the disturbed periods, 36 years for the control, and 4 years for the damage periods.

3.1.2. Discharge Duration Curve

A discharge duration curve was created to replace the daily discharge volume data for 365 days in larger order, as shown in figure 3. Higher and lower daily discharge volumes are shown on the left and right sides of the curve, respectively. The 365 days of daily data shown on the curve were divided into categories separated at the 95th-largest, 185th-largest, 275th-largest, and 355th-largest daily values, representing "plentiful," "ordinary," "low," and "drought" water discharge, respectively. In this study, the annual discharge volume was obtained over one water year, defined as 1 January to 31 December.

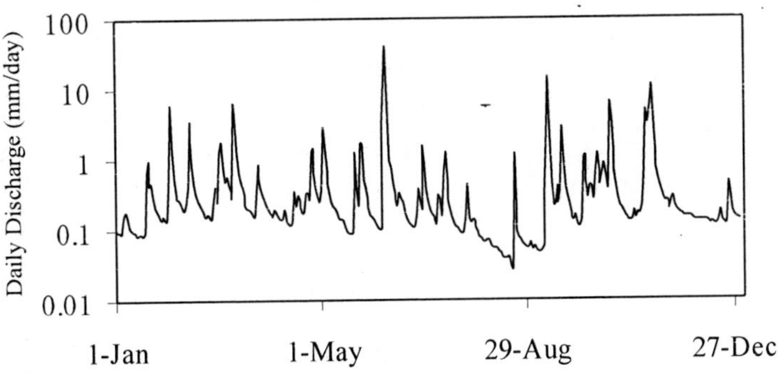

(A)

Figure 3. (Continues)

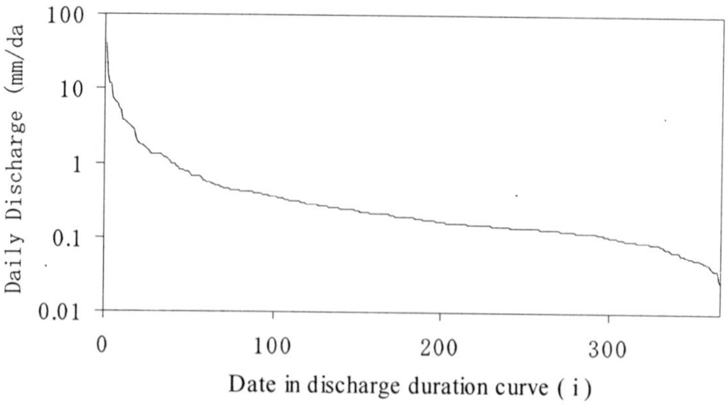

(B)

Figure 3. Sample of daily discharge volume and discharge duration curve. A) Daily discharge volume; B) Discharge duration curve.

3.2. Results and Discussion

3.2.1. Comparison of Discharge Duration Curves

Figure 4 shows the plentiful, ordinary, low, and drought water discharges of the two catchments. The X- and Y-axes show the discharge volume from the control (KT) and disturbed (MN) catchments, respectively. As noted above, the disturbed, control, and damage periods included 19, 36, and 4 years, respectively; these are represented in the figures by white squares, black diamonds, and asterisks, respectively. Regression lines expressed by Eq. (2) were obtained with high correlation for the disturbed and control periods, respectively:

$$Q_{MN}(year, i) \quad a_{p,i}\, Q_{KT}(year, i) + b_{p,i} \tag{2}$$

where Q (year, i) indicates the ith largest daily discharge volume (mm day^{-1}) in a certain year, and $a_{p,i}$, $b_{p,i}$ show the regression coefficient and regression constant, respectively. The disturbed and control periods are indicated by p = d and p = c, respectively.

Next, $Q_{MNcal}(year, i)$ was calculated with Eq. (3) by substituting the measured $Q_{KT}(year, i)$ for the disturbed period into Eq. (2) obtained for the control period. $Q_{MNcal}(year, i)$ can be regarded as the estimated daily discharge volume in the case that the forest was not disturbed in the disturbed (MN) catchment:

$$Q_{MNcal}(year, i) = a_{p,I}\, Q_{KTobs}(year, i) + b_{p,i} \tag{3}$$

In figure 4, white squares plotted left of and above, and right of and below, the Eq. (2) regression line for the control period indicate increases and decreases, respectively, of the daily discharge volume from the disturbed catchment relative to that in the control catchment. All white squares are plotted above and to the left of this line in figure 4, indicating that forest disturbance increased daily discharge volumes. Maita and Suzuki (2008) reported the results of a similar analysis on the effect of clear cutting a coniferous evergreen forest in the Fukuroyamasawa experimental catchment in Tokyo University Forest, Chiba, Japan. They reported that the discharge volume was relatively increased in the clear-cut catchment after cutting. Our results agree with their findings.

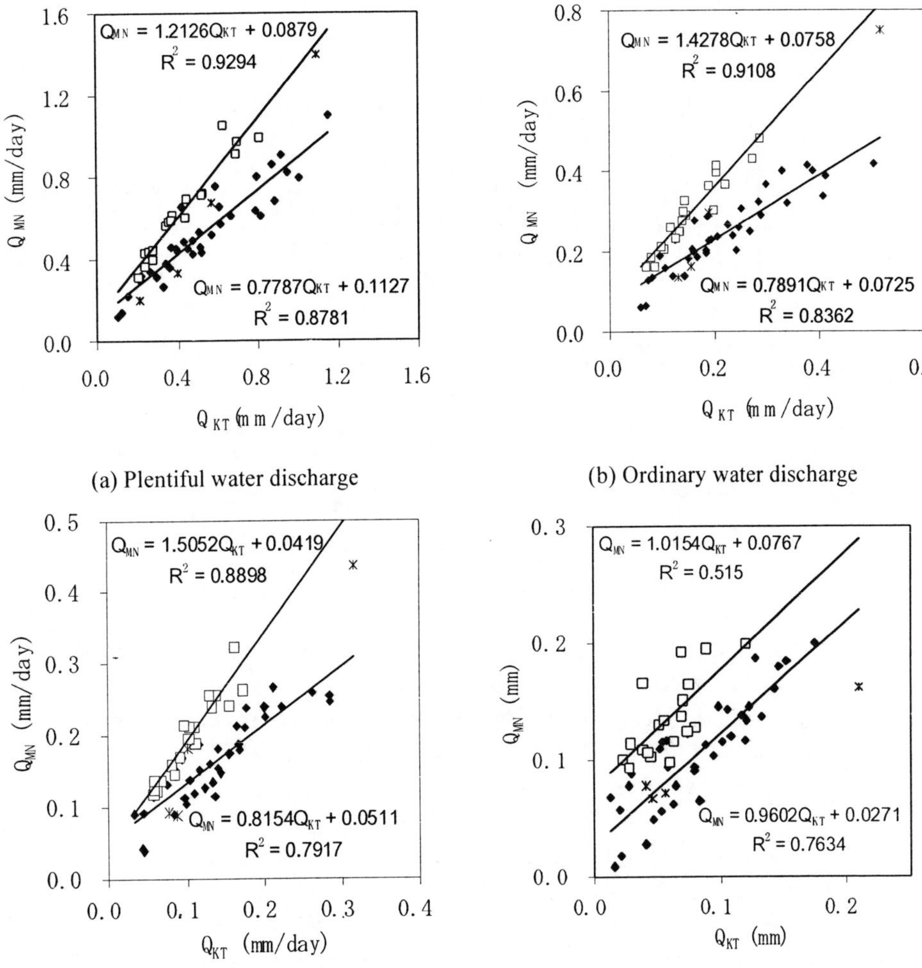

(a) Plentiful water discharge

(b) Ordinary water discharge

(c) Low water discharge

(d) Drought water discharge

Figure 4. Relationship between control and distributed watersheds in control and disturbed periods. Thick kine: Eq.(3); Black diamond: Control period, White square: Disturbed period, Astarisk: Damage period.

3.2.2. Increased Discharge Volume with Disappearance of the Forest

Using Eq. (4), we calculated ΔQ_m(year, i) to estimate the increased discharge volume caused by disappearance of the forest in the MN catchment during the disturbed period:

$$\Delta Q_{MN}(\text{year, i}) = Q_{MNobs}(\text{year, i}) - Q_{MNcal}(\text{year, i}) \quad (4)$$

In the above equation, Q_{MNobs}(year, i) represents observed discharge volume from the disturbed (MN) catchment in the disturbed period.

Table 2 shows average ΔQ_{MN}(year, i) in each disturbed period caused by forest fire and pine wilt disease and the results of Maita and Suzuki (2008). As indicated in the table, ΔQ_{MN}(year, i) in the Tatsunokuchi-yama forest was much smaller than that in the Fukuroyamasawa forest. These differences were likely caused by differences in annual precipitation and tree volumes in the control periods between the two forests.

Table 2. Increased discharge volume caused by forest disturbance

		1960-1966 (Forest fire)	1981-1994 (Pine Disease)	Maita & Suzuki (2008)
Plentiful	ΔQ_m (year,i)	0.227	0.158	0.71
	Q_{mobs} (year,i)	0.707	0.566	
Ordinary	ΔQ_m (year,i)	0.108	0.098	0.33
	Q_{mobs} (year,i)	0.32	0.278	
Low	ΔQ_m (year,i)	0.066	0.059	0.20
	Q_{mobs} (year,i)	0.213	0.184	
Drought	ΔQ_m (year,i)	0.048	0.056	0.12
	Q_{mobs} (year,i)	0.136	0.131	

Bosch and Hewlett (1982) reported that the amount of increase in discharge volume caused by forest change depends on the precipitation amount. The Tatsunokuchi-yama forest receives much less annual precipitation (approximately 1,200 mm) than the Fukuroyamasawa forest (approximately 2,170 mm; Maita and Suzuki, 2008).

Further, the dominant species in Fukuroyamasawa before clear cutting were sugi (*Cryptomeria japonica*) and hinoki, with a tree volume of 500 $m^3 ha^{-1}$ (Maita and Suzuki, 2008). At Tatsunokuchi-yama just before the disturbances by fire in 1959 and pine wilt disease in 1978, tree volumes were reported to be 0.9 m^3 ha^{-1} and 43.4 m^3 ha^{-1}, respectively (Fujieda and Abe, 1982). Thus, we can assume that the forest at Tatsunokuchi-yama was in poorer condition and likely transpired less water than the forest at Fukurosawa. The smaller ΔQ_{MN}(year,i) at Tatsunokuchi-yama than at Fukuroyamasawa can therefore be explained by the lower precipitation and poorer forest condition of the former.

Figure 5 shows the average ΔQ_{MN}(year, i) for disturbed period and standard deviation for every i. The average ΔQ_{MN}(year, i) decreased as i increased. The fine lines indicate the standard deviation. The maximum and minimum average ΔQ_{MN}(year, i) were 4.114 mm day^{-1} at i=4 and 0.050 mm day^{-1}at i=362, respectively. Figure 6 shows the measured Q_{MN}(year, i) divided into Q_{MNcal}(year, i) and ΔQ_{MN}(year, i), which suggest the estimated discharge from MN under normal conditions and forest decline, respectively. The upper light gray and lower dark gray areas indicate the shares of ΔQ_{MN}(year, i) and Q_{MNcal}(year, i), respectively. Generally, the share of ΔQ_{MN}(year, i) increased with i, with shares of 15, 35, and 43% for i values of 1–40, 140–320, and 365.

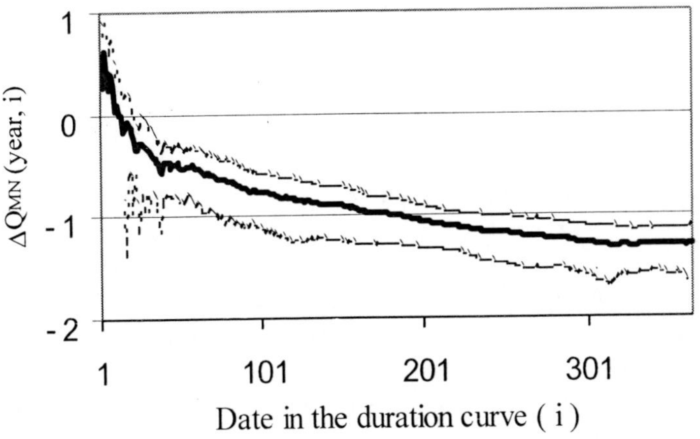

Figure 5. Averaged ΔQ and standard deviation in disturbed period. Thick line: Averaged ΔQ, Thin line: Standard deviation.

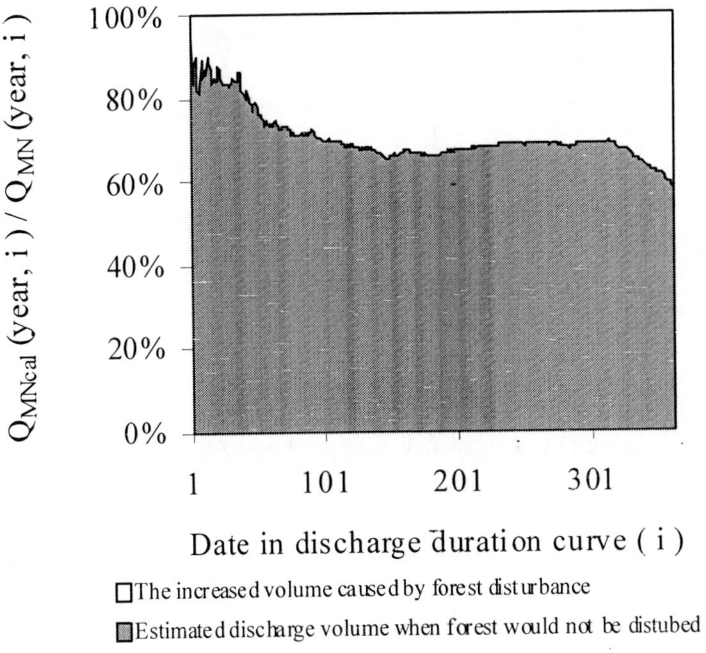

Figure 6. The shares of ΔQ and Q_{calMN} in Q_{obsMN}.

4. Estimating the Effects of Various Forest Disturbances in the MN Catchment

4.1. Method: Defining the New Index

Water discharge levels, i.e., plentiful ($i = 95$), ordinary ($i = 185$), low ($i = 275$), and drought ($i = 355$), were analyzed among the discharge duration curves. The new index distinguished between plentiful, ordinary, and low water discharge levels, which had similar values of b(t, i) and b(c, i), and the drought water discharge level with values of a(t, 355) and a(c, 355). Figure 7(a) illustrates the newly defined index for drought water discharge. Lines C and D are the regression lines for the control and disturbed periods shown in figure 4(d), respectively. The values were determined such that a(t, 355) = 0.9602 and a(c, 355) = 1.0154, with an average value of 0.9878. Line Y, which passes through Point Q, indicates Q_{KT}(year, 355). Eq.(5), given below, was used to determine a slope of 0.9878 for Q_{MN}(year, 355) in any year:

$$Q_{MN}(\text{year}, 355) = 0.9878\ Q_{KT}(\text{year}, 355) + b(\text{year}) \quad (5)$$

For values of b(year) closer to 0.0271 and 0.0767, Point Q is plotted closer to Line C and Line D, respectively. This relationship was standardized in Eq. (6):

$$I(\text{year}, 355) = 2(b(\text{year}) - 0.0271)/(0.0767 - 0.0271) - 1 \quad (6)$$

Values of I(year, 355) equal to -1.0, 1.0, and 0 indicate that Point Q falls on Line C, on Line D, and midway between these two lines, respectively.

Figure 7(b) shows the definitions of the new index for plentiful, ordinary, and low water discharges. Lines C and D indicate the regression lines for the control and disturbed periods in figure 4(d), respectively. As shown in figures 4(a), (b), and (c), Line C and Line D intersect at Point P. The coordinates of Point P were (0.05716, 0.157207), (-0.00517, 0.06842), and (0.01334, 0.06198) for plentiful, ordinary, and low water discharge, respectively. Line Y, connecting Points Q and P, was obtained for every year. Thus, the relative distances between Point Q and Lines C and D can be estimated by comparing the slopes of Lines D, C, and Y. This relationship was calculated with Eq. (7) as follows:

$$I(\text{year}, i) = 2(\theta_{QPH} - \theta_{NPH})/(\theta_{DPH} - \theta_{NPH}) - 1 \quad (7)$$

where θ_{QPH}, θ_{NPH}, and θ_{DPH} indicate the angles in degrees between Line Y, Line C, and Line D, respectively, and Line H, which is parallel to the X-axis. Values of I(year, i) equal to -1.0, 1.0, and 0 indicate that Point Q falls on Line C, on Line D, and midway between these two lines, respectively.

I(year, i) calculated with Eqs. (6) and (7) is the new index proposed in this study.

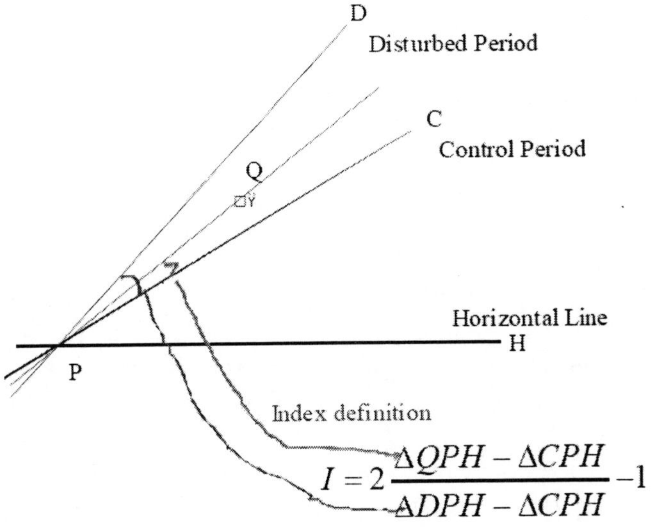

(a) New index for plentiful, ordinary and low water discharge

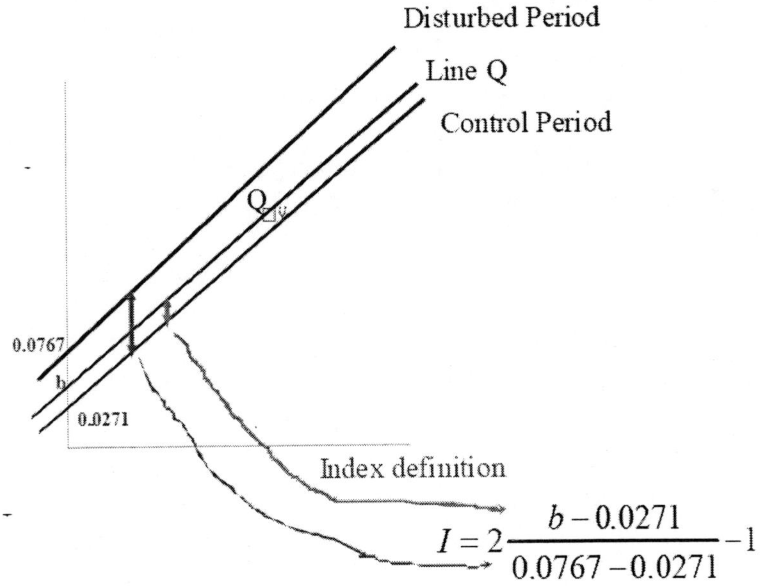

(b) New index for drought discharge

Figure 7. Definition of new index I(year,i).

4.2. Results and Discussion

4.2.1. Fluctuation of I(Year, I) and Forest Information

Figure 8 shows the inter-annual fluctuations of I(year, i) from 1937 to 2003. If I(year, i) is 0, the white square in figure 4 is plotted midway between Line C and Line D. Positive and negative values of I(year, i) would place the white square in figure 4 nearer to Line D and to Line C, respectively. Almost all I(year, i) values are negative in the control period. When a white square is plotted on Line C, I(year, i) is around -1. Thus, a value of I(year, i) between -1 and 0 also means that ΔQ_{MN}(year,i is positive. However, the deviation of daily discharge volumes from both catchments in the control period for Line C was judged to have a larger effect on I(year, i) than the effect of forest disturbance when I(year, i) is between -1 and 0. Thus the effect of forest disturbance can be recognized when I(year, i) is larger than 0 in this study.

Almost all values of I(year, i) were positive in the disturbed periods of 1960–1966 and 1981–1994. Thus, this indicates that almost all of the daily discharge volumes shown in figure 4 from the disturbed catchment increased in the disturbed periods. Strictly speaking, the fluctuation trend of only I(year, 355) for drought water discharge is different from I(year, i) for plentiful, ordinary, and low water discharge. For example, I(year, 355) was negative even though other I(year, i) values were positive in 1959 (just before the disturbed period) and in 1980 (the last year of the second damage period). This result may have been caused by the definition of a water year used in this study. By using 1 January to 31 December as a water year, one discharge duration curve could include data from two drought seasons, namely from 1 January into spring and from autumn to 31 December. However, only the data from 1 January to spring in a certain discharge duration curve would be affected by the precipitation in the preceding year. Thus, some drought water discharge data would be affected by the amount of precipitation in the same water year, while others would be affected by the amount of precipitation in the preceding water year. Inaba et al. (2007) defined the water year for the Tatsunokuchi-yama forest as 1 May to 30 April to obtain a figure like figure 4 in this study; they found a higher correlation coefficient for the drought water discharge. Their discharge duration curve included data from only one drought season and excluded the effect of the precipitation amount in the preceding water year on drought water discharge data.

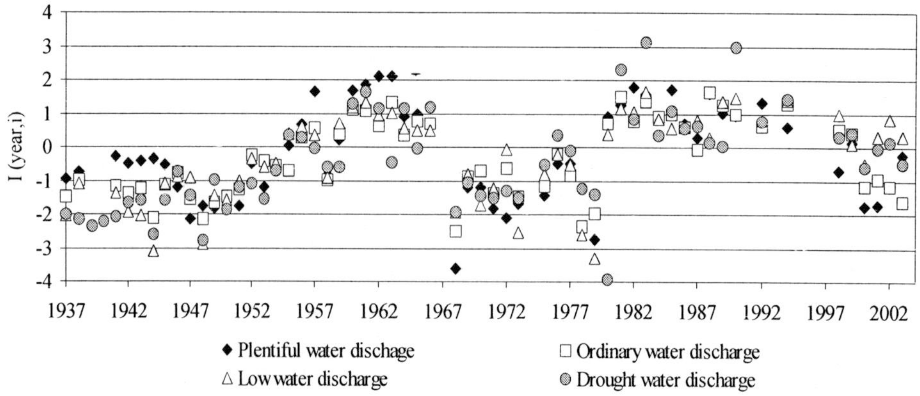

Figure 8. Fluctuation of new index I(year, i).

Positive I(year, i) values were found in the control period. Most I(year, i) were positive in the period 1955–1957. The only negative value in this period I(1957, 355) was –0.04, which was very close to 0. Brush cutting took place in the MN catchment in 1954, 1955, and 1957. Each time, 4 to 8 ha of brush were cut. Moreover, 14.7 and 4.8 ha of understory trees were removed in 1954-1957 (Nakano, 1971). Although the MN catchment forest often received management treatments in those years, the stem volume in MN at that time is unknown. However, the stem volume in KT was reported to be very small at only 0.9 $m^3 m^{-3}$ (Fujieda et al., 1982). At this time, the vegetation in the MN and KT catchments was natural regrowth following clear cutting in 1944–1947. Thus, the stem volume in the MN catchment was probably small, like that in the KT catchment. Thus, even small disturbance such as brush cutting and understory removal would have deeply impacted the poor vegetation with small stem volume in the MN catchment from 1955–1957.

4.2.2. Forest Regrowth and Fluctuations of I(Year, I)

Approximately half of I(year, i) values were positive in the period 1998–2002, which followed a disturbed period of pine wilt disease. These years were categorized in the control period because ΔQ was negative; however, the damage caused by pine wilt disease apparently remained even 22 years after the last year of the disturbed period (1980). In contrast, I(year, i) became negative 11 years after the forest fire damage period. This difference may have been caused by differences in treatment following the disturbance. After the pine wilt period, vegetation was allowed to re-grow naturally. In contrast, after the forest fire, a plantation of black pine was planted in March 1960, which would have helped vegetation and water discharge recover more quickly.

The decreasing trend in I(year, i) from 1998–2002 suggests that the effect of pine wilt disease decreases only gradually. Further, the finding that I(year, 95) and I(year, 185) became negative earlier than I(year, 275) and I(year, 355) suggests that recovery from pine wilt disease is more rapid at times of high water than in times of low water.

The effects of smaller disturbances were not recognized in the water discharge; these smaller events included a forest fire in 1974, partial clear cutting in 1977, brush cutting and hinoki planting in 1982, and hinoki thinning in 1998 and 2001. At these times, there was either very small or unclear increase of I(year, i).

REFERENCES

Abe T. & Tani M. (1985) Streamflow changes after killing of pine trees by the pine-wood nematode. *Journal of the Japanese Forestry Society*. 67. 261-270.

Bosch J. M. & Hewlett J. D. (1982) A review of catchment experiments to determine the effect of vegetation changes on water yield and evapotranspiration. *Journal of Hydrology*. 55:3-23.

Fahey B. & Jackson R. (1997) Hydrological impacts of converting native forests and grasslands to pine plantations, South Island, New Zealand. *Agricultural and Forest Meteorology*. 84. 69-82.

Fujieda M. & Abe T. (1982) Afforestation on streamfrow on Tatsunokuchiyama Experimental Watershed. *Bulletin of Forestry & Forest Products Research Institute.* No317. 113-138.

Hattori S. (1994) Water balance in forest in small rainfall region. *Journal of Japanese Society of Revegetation Technology.* 19. 296-302.

Inaba N., Kondo K., Numamoto S. & Hayashi S. (2007) Influence of the definition of water-year period on discharge-duration analysis focused on low flow: In the case of the Tatsunokuchi-yama Experimental Watershed. *Journal of Japanese Forestry Society.* 89. 412-415.

Kobayashi K. Okuda M. & Hosoda R. (1981) Tree damage by pine wilt disease at Tatsunokuchi-yama*. *Annual Report of the Kansai Branch, Forestry and Forest Products Research Institute.* 22, 63-69.

Maita E. & Suzuki M. (2008) The effect of forest cutting on flow duration curve of mountainous small watershed: Evaluation of the influence of vegetation on inter watershed variation in flow duration curve. *Journal of Japanese Forestry Society.* 90. 36-45.

Nakano H. (1971) Effect on streamflow of forest cutting and change in regrowth on cut-over area. *Bulletin of Governmetal Forestry Experimental Station.* No240. 1-251.

In: Environmentalism: Environmental Strategies...
Editors: E. Kahraman and A. Baig, pp.183-190
ISBN: 978-1-60741-634-0

Chapter 9

ARE ENVIRONMENTAL FACTORS CAPITALIZED INTO THE NOMINAL MEDIAN PRICE OF SINGLE-FAMILY HOMES?

***Richard J. Cebula,*[1] *Luther Lawson*[2] *and Usha Nair-Reichert*[3]**

[1] Armstrong Atlantic State University, GA, USA
[2] University of North Carolina, Wilmington, NC, USA
[3] Georgia Institute of Technology, GA, USA

ABSTRACT

Using a semi-log model specification, this empirical chapter investigates determinants of nominal median single-family home price differentials for the year 2005 using state-level data. Although the literature on geographic living-cost differentials is well developed, the literature on geographic housing price differentials is much less so. Given the relatively large impact of housing prices on overall living costs, this research seeks to address this issue and shed light on specific factors influencing the nominal median price of single-family homes across states. Particular emphasis is placed on the capitalization of environmental factors such as environmental pollution in the form of toxic chemical releases and favorable climatic conditions into housing prices. As control variables, several economic variables and a measure of fiscal surplus are also included in the analysis. The semi-log results imply that the natural log of the nominal median price of single-family homes is negatively a function of toxic chemical releases. In addition, it is positively a function of measures of both warm climate and sunny climate. Furthermore, the natural log of the median price of single-family homes is an increasing function of median family income, the percentage employment growth rate, the extent of unionization, and the magnitude of fiscal surplus.

INTRODUCTION

The determinants of geographic living-cost differentials (L-CDs) in the U.S has attracted the interest of a large number of researchers, including Cebula (1980, 1989), Cebula & Todd (2004), Cobas (1978), Ostrosky (1983), McMahon (1991), Nord (2000), and Kurre (2003). These studies focus on states, metropolitan areas, and counties. The study of geographic L-CDs is of interest given that such differentials have consistently been found to be significant in explaining geographic mobility in the U.S. (Renas, 1978, 1980, 1983; Cebula, 1978, 1993; Cebula & Alexander, 2006; Ashby, 2007).

Less well developed is the literature on interstate housing-price differentials. Based on the most recent consumer price index methodology, expenditures on residential housing constitute approximately 40 percent of total expenditures by consumers. Furthermore, as Ashby (2007, p. 686) observes, "...housing...is the main driver of cost-of-living differences between states." Moreover, given the ongoing turbulence in housing markets nationwide, an investigation of interstate differentials in housing prices is of interest in its own right and may shed light on the factors accounting for the substantial variations in housing market performance across the country. Accordingly, this empirical chapter estimates a semi-log model to focus on the single most important component of L-CDs, that of housing prices, by investigating determinants of nominal median prices on single-family homes across states for the year 2005. Moreover, given the relevance of environmental issues in the modern economy and on the political stage, this investigation of housing price determinants places special emphasis on the degree to which the market reacts to environmental externalities. In particular, the chapter will significantly focus on the degree (if any) environmental factors are capitalized by the market into housing prices.

THE SEMI-LOG APPROACH

As illustrated in the comprehensive survey articles by Boyle & Kiel (2001) and Sirmans, Macpherson & Zeitz (2005), hedonic pricing models have been applied by researchers to a wide variety of circumstances in recent years. For example, the hedonics framework has been applied to the issue of flood hazard and hurricane damage (Bin & Polasky, 2004; Harrison, Smersh & Schwartz, 2001), the impact of air quality, environmental, and safety issues (Couton, Gardes & Thepauti, 1996; Decker, Nielsen & Sindt, 2005; Dotzour, 1997; Garrod & Willis, 1992; Graves, Murdock & Thayer, 1988; Levesque, 1994; Murdoch, Singh & Thayer, 1993; Neill, Hassenzahl & Assane, 2007; Netusil, 2005; Smith & Huang, 1993), treating land as a factor of production (Palmquist, 1989), and the issue of over-education and under-education in a dynamic labor market (McMillen, Seaman & Singell, 2007).

The technique also is widely used to explain and understand housing-price determinants. Indeed, going further, hedonic pricing models have been even used to assess the impacts of historic district designation on property values (Coffin, 1989; Ford, 1989; Asabere & Huffman, 1994; Asabere, Huffman & Mehdian, 1994; Clark & Herrin, 1997; Coulson & Leichenko, 2001; Leichenko, Coulson & Listokin, 2001). The present study adopts this resilient technique to estimate whether and to what extent environmental factors such as pollution in the form of toxic chemical emissions are actually capitalized into single-family house prices. The data, unlike most such applications, are state-level.

THE FRAMEWORK

Consistent in principle with the research on L-CDs, the perspective underlying this analysis of housing prices is that factors affecting housing supply and housing demand, including the relative attractiveness of a geographic area in terms of amenities and negative amenities, influence those housing prices. Factors that directly elevate housing demand or represent amenities and thus indirectly raise housing demand through capitalization tend to elevate overall housing prices in an area. By contrast, factors tending directly to increase the relative supply of housing in an area or to indirectly reduce the demand for (desirability of) of housing in an area through capitalization of negative externalities (negative amenities) lead to lower housing prices in that area. In this empirical analysis, the dependent variable, $PHOME_j$, is defined as the nominal median price of single-family homes in state j in 2005. Its determination is the focal point of this study.

It is expected that the greater the net percentage employment growth rate (EMPGRj) in an area, the greater the growth in the overall demand for housing in the area, thus yielding higher housing prices, *ceteris paribus*. In other words, greater net growth in the employment base, whether attributable to endogenous growth or net in-migration of workers, would contribute to the demand for available housing and land and create upward pressure on housing prices. It also is expected that, assuming single-family homes are a "normal good," the higher the median family income (MFIj) an area, the higher the demand for housing, including that for single-family homes, *ceteris paribus*, and hence the higher the price of those homes. From a different perspective, the greater the percentage of the labor force that is unionized (UNIONj) in an area, the greater the unit labor costs (Cebula, 1980; 1989) and hence the greater the unit construction costs and, as a result, the higher the prices on single-family homes, *ceteris paribus.*

Next, the role of climate has been shown to be a significant determinant of migration (Cebula, 1978; Renas, 1978, 1980, 1983; Ashby, 2007). In this analysis, climate is in part proxied by the average January temperature (JANTEMPj*)*. Clearly, this variable is higher in warmer climates, reflecting the desirable feature (amenity) of warmer temperatures that presumably may be *capitalized* into housing prices. Accordingly, PHOMEj is expected to be positively related to JANTEMPj, *ceteris paribus.* Climate is also partly proxied by sunshine, i.e., the percentage of daylight that, on the average, is characterized by sunshine (as opposed to clouds or inclement weather), SUNj. Arguably, like its complement JANTEMPj, the value of SUNj is hypothesized to be positively capitalized into housing prices, so that both of these "favorable" climatic conditions of the environment are expected to raise home prices. In addition, the effect of toxic chemical releases (TOXICj) on PHOMEj is considered. This variable reflects pollution conditions in each state, and as an undesirable feature, (negative externality) is expected to be negatively *capitalized* into housing prices. In particular, higher levels of toxic chemical releases in an area are expected to be considered a negative amenity for would-be home buyers and are thusly expected to indirectly reduce the demand for and hence the median price of single-family homes in that area, *ceteris paribus*..

The remaining variable is that of "fiscal surplus," measured in the case of the present study as the excess of the value of annual per pupil expenditures on *public* primary plus secondary education minus per capita property taxes (FISCSURPj). According to Tiebout (1956, p. 418), "...the consumer-voter may be viewed as picking that community which best satisfies his preferences for public goods...the consumer-voter moves to that community whose local government best satisfies his set of preferences." As Tullock (1971, p. 917) further observes, this hypothesis effectively can be extended such that it holds that, *ceteris paribus*, the "...individual deciding where to live will take into account the private effects upon himself of the bundle of government services and taxes..." Thus, Tullock (1971) more explicitly emphasizes that the consumer-voter evaluates *both* the government goods and services *and* the tax burden at locations of choice. Based on these two studies, it is expected that fiscal surplus, as defined above, will be capitalized (positively) into housing prices, *ceteris paribus*.

EMPIRICAL MODEL AND FINDINGS

Based on the eclectic model developed above, in this chapter, two versions of the following basic reduced-form equation are estimated:

$$PHOMEj = a + b\,MFIj + c\,EMPGRj + d\,UNIONj + e\,JANTEMPj + f\,SUNj + g\,TOXICj + h\,FISCSURPj + u \qquad (1)$$

where:

PHOMEj = the nominal median single-family home price in state j in 2005;
a = constant;
MFIj = median family income in state j in 2004;
EMPGRj = the percentage net growth rate of employment in state j, 2000-2004;
UNIONj = the percentage of the civilian labor force that was unionized in state j in 2004;
JANTEMPj = average high January temperature in degrees Fahrenheit in state j (average for period 1971-2000);
SUNj = the average percentage of daylight in state j in each year during which there is sunshine (average for the period 1974-2000);
TOXICj = toxic chemical releases per square mile into the air, water, or land of state j, in thousands of pounds, 2003;
FISCSURPj = the fiscal surplus in state j in 2004; and
u = stochastic error term.

Based on the arguments provided in Section II above, it is expected that:

$$b > 0,\ c > 0,\ d > 0,\ e > 0,\ f > 0,\ g < 0,\ h > 0 \qquad (2)$$

The data sources are, as follows:

EMPGRj: U.S. Census Bureau (2001, Table 572) and U.S. Census Bureau (2006, Tables 17, 619); UNIONj, JANTEMPj, SUNj, TOXICj, FISCSURPj: U.S. Census Bureau (2006, Tables, 27, 377, 368, 369, 371, 295, 430, 17); PHOMEj: http://sdc.state.nc.us/.

Estimating equation (1) in semi-log form by OLS, adopting the White (1980) heteroskedasticity correction, yields the following:

lnPHOMEj = 8.76 + 0.0268 EMPGRj + 0.00003 MFIj + 0.022 UNIONj + 0.0098 JANTEMPj

(+3.81) (+10.49) (+3.31) (+4.46)

+ 0.0081 SUNj - 0.121 TOXICj + 0.00215 FISCSURPj,

(+2.68) (-4.12) (+1.85)

$R^2 = 0.85$, adj$R^2 = 0.83$, F = 35.15 (3)

where terms in parentheses are t-values. In equation (3), all seven of the estimated coefficients exhibit the expected signs, with six being statistically significant at the one percent level, and the remaining one (FISCSURPj) being statistically significant at beyond the ten percent level. The coefficient of determination is 0.85, so that the model explains more five-sixths of the variation in the nominal median price of single-family homes across states for the year 2005.

In this estimate, which uses state-level data, the estimated coefficient on the EMPGR variable is positive and statistically significant at the one percent level. Thus, there is compelling evidence that the greater the percentage growth rate of employment, the higher the median price of single-family homes, presumably because of the increased demand for same resulting from the employment growth. Next, the estimated coefficient on the MFI variable is also positive and statistically significant at the one percent level, implying that the higher the median family income, the higher the demand for single-family homes and hence the higher the price of such homes. Single-family homes would appear to be normal goods. Similarly, the coefficient on the variable UNION is positive and significant at the one percent level, implying that the price of single-family homes is positively a function of the degree of unionization. This result may reflect higher labor costs associated with a greater degree of labor force unionization as argued above and in Cebula (1980; 1989) and Ostrosky (1983).

Turning now that factors that may be relevant to the issue at hand but *strictly* because they are capitalized into housing prices, we first consider the JANTEMP variable. The estimated coefficient on this variable is positive and statistically significant at the one percent level, implying that the attribute/amenity of warm climate is capitalized into the price of single-family homes. The estimated coefficient on the SUN variable also is positive and statistically significant at the one percent level, suggestive that the value of sunshine among various climatic traits is also capitalized into housing prices. Next, the estimated coefficient on the TOXIC variable is negative and significant at the one percent level, implying that greater degrees of toxic chemical emissions are capitalized (negatively) into the prices of single-family homes. Thus, negative externalities have real-world market implications. The desirability of homes in areas with higher toxic emissions is clearly reduced as would-be homebuyers discount market prices to reflect the negative amenity value of such pollution. Finally, the estimated coefficient of the Tiebout-Tullock "fiscal surplus" variable, PROPTX, is positive (as hypothesized) but statistically significant at only the seven percent level. This finding is weakly consistent with arguments central to the classic papers by Tiebout (1956) and Tullock (1971).

Given the weakness of the FISCSURP variable in the semi-log estimate in equation (3), we re-estimate equation (1) but without this variable integrated into the system. This estimate constitutes the second version of the basic model. The OLS, White (1980)- corrected estimate is given by equation (4):

lnPHOMEj = 8.92 + 0.0256 EMPGRj + 0.00003 MFIj + 0.0237 UNIONj + 0.0107 JANTEMPj

(+3.63) (+10.36) (+3.63) (+5.07)

+ 0.0065 SUNj - 0.121 TOXICj, R^2 = 0.85, adjR^2 = 0.83, F = 40.46 (4)

(+2.22) (-4.10)

Clearly, these results offer no surprises. Indeed, except for the case of the SUN variable (whose coefficient falls from statistically significant at the one percent level to significant at the three percent level), the robustness of the model is modestly improved.

Conclusion

This empirical chapter finds that, for the year 2005, the nominal median price of a single-family home was an increasing function of the net percentage growth rate of employment, median family income, the degree of unionization, the mean January temperature, and degree of sunshine in the climate, and, less compellingly, fiscal surplus. In addition, the nominal median price of a single-family home was found to be a decreasing function of pollution (as measured by toxic chemical releases). Thus, it is inferred strongly that environmental considerations are formally capitalized into housing prices.

References

Asabere, P.K., & Huffman, F. E., (1994). Historic Designation and Residential Market Values. *Appraisal Journal*, 62, 396-401.

Asabere, P.K., Huffman, F. E., & Mehdian, S. (1994). The Adverse Impacts of Local Historic Designation: The Case of Small Apartment Buildings in Philadelphia. *Journal of Real Estate Finance and Economics*, 8, 225-234.

Ashby, N.J. (2007). Economic Freedom and Migration Flows between U.S. States. *Southern Economic Journal*, 73, 677-697.

Bin, O., & Polasky, S. (2004). Effects of Flood Hazards on Property Values: Evidence before and after Hurricane Floyd. *Land Economics*, 80, 490-500.

Boyle, M.A., & Kiel, K.A. (2001). A Survey of House Price Hedonic Studies of the Impact of Environmental Externalities. *Journal of Real Estate Literature*, 9, 117-144.

Cebula, R.J. (1978). An Empirical Note on the Tiebout-Tullock Hypothesis. *Quarterly Journal of Economics*, 92, 705-711.

Cebula, R.J. (1980). An Empirical Note on Determinants of Geographic Living Cost Differentials. *Land Economics*, 56, 477-481.

Cebula, R.J. (1989). The Analysis of Geographic Living Cost Differentials: A Brief Empirical Note. *Land Economics*, 65, 64-67.

Cebula, R.J. (1993). The Impact of Living Costs on Geographic Mobility. *Quarterly Review of Economics and Finance*, 33, 101-105.

Cebula, R.J., & Alexander, G. (2006). Determinants of Net Interstate Migration, 2000-2004. *Journal of Regional Analysis and Policy*, 36, 115-127.

Cebula, R.J., & Todd, S. (2004). An Empirical Note on Determinants of Geographic Living-Cost Differentials for Counties in the State of Florida, 2003. *Review of Regional Studies*, 34, 112-119.

Clark, D. E., & Herrin, W. E. (1997). Historical Preservation Districts and Home Sales: Evidence from the Sacramento Housing Market. *Review of Regional Studies*, 27, 29-48.

Cobas, J. A. (1978). A Method to Estimate BLS Family Budgets for All Standard Metropolitan Statistical Areas. *Social Science Quarterly*, 59, 546-552.

Coffin, D. A. (1989). The Impact of Historic Districts on Residential Property Values. *Eastern Economic Journal,* 15, 221-228.

Coulson, N. E., & Leichenko, R. M. (2001). The Internal and External Impact of Historical Designation on Property Values. *Journal of Real Estate Finance and Economics*, 23, 113-124.

Couton, C., Gardes, F., & Thepauti, Y. (1996). Hedonic Prices for Environmental and Safety Characteristics and the Akerlof Effect in the French Car Market. *Applied Economics Letters*, 3, 435-440.

Decker, C.S., Nielsen, D.A., & Sindt, R.P. (2005). Is Pollution a Homogeneous Determinant of Value? *The Appraisal Journal*, 73, 83-196.

Dotzour, M.G. (1997). Groundwater Contamination and Residential Property Values. *The Appraisal Journal*, 65, 279-287.

Dubin, R. (1998). Predicting House Prices Using Multiple Listings Data. *Journal of Real Estate Finance and Economics*, 17, 35-60.

Ford, D A. (1989). The Effect of Historic District Designation on Single-Family Home Prices. *AREUEA Journal*, 17, 353-362.

Garrod, G., & Willis, K.G. (1992). The Environmental Economic Impact of Woodland: A Two State Hedonic Price Model of the Amenity Value of Forestry in Britain. *Applied Economics*, 24, 715-728.

Graves, P., Murdoch, J., & Thayer, M. (1988). The Robustness of Hedonic Price Estimation: Urban Air Quality. *Land Economics*, 64, 221-233.

Harrison, D.M., Smersh, G.T., & Schwartz, A.L. (2002). Environmental Determinants of Housing Prices: The Impact of Flood Zone Status. *Journal of Real Estate Research*, 21, 3-20.

Kurre, J.A. (2003). Is the Cost of Living Less in Rural Areas? *International Regional Science Review,* 26, 86-116.

Laurice, J., & Bhattacharya, R. (2005). Prediction Performance of a Hedonic Pricing Model for Housing. *The Appraisal Journal*, 73, 198-209.

Leichenko, R. M., Coulson, N. E., & Listokin, D. (2001). Historic Preservation and Residential Property Values: An Analysis of Texas Cities. *Urban Studies*, 38, 173-187.

Levesque, T.J. (1994). Modeling the Effects of Airport Noise on Residential Housing Markets. *Journal of Transport Economics and Policy*, 28, 199-210.

McMahon, W.W. (1991). Geographical Cost of Living Differences: An Update. *AREUEA Journal,* 19, 426-450.

McMillen, D.P., Seaman, P.T., & Singell, L.D. (2007). A Mismatch Made in Heaven: A Hedonic Analysis of Overeducation and Undereducation. *Southern Economic Journal*, 73, 901-930.

Neill, H. R., Hassenzahl, D. M., & Assane, D. D. (2007). Estimating the Effect of Air Quality: Spatial versus Traditional Hedonic Price Models. *Southern Economic Journal*, 73, 1088-1111.

Netusil, N. R. (2005). The Effect of Environmental Zoning and Amenities on Property Values: Portland, Oregon. *Land Economics*, 81, 227-246.

Nord, M. (2000). Does It Cost Less to Live in Rural Areas? Evidence from New Data on Food Security and Hunger. *Rural Sociology*, 65, 104-125.

North Carolina State Data Center, 2005: http://sdc.state.nc.us.

Ostrosky, A.L.(1983). Determinants of Geographic Living Cost Differentials: Comment. *Land Economics*, 59, 212-213.

Palmquist, R. B. (1989). Land as a Differentiated Factor of Production: A Hedonic Model and Its Implications for Welfare Measurement. *Land Economics*, 65, 23-28.

Putnam, R. (2000). *Bowling Alone*. New York: Simon & Schuster.

Renas, S. (1978). The Cost of Living, Labor Market Opportunities, and the Migration Decision. *Annals of Regional Science*, 15, 95-104.

Renas, S. (1980). An Empirical Note on the Tiebout-Tullock Hypothesis: Comment. *Quarterly Journal of Economics*, 94, 619-623.

Renas, S. (1983). The Cost of Living, Labor Market Opportunities, and the Migration Decision: More on Problems of Misspecification and Aggregation Bias. *Annals of Regional Science*, 17, 98-110.

Sirmans, G. S., Macpherson, D. A., & Zeitz, E. N. (2005). The Composition of Hedonic Pricing Models," *Journal of Real Estate Literature*, 13, 3-43.

Smith, V.K., & Huang, J.C. (1993). Hedonic Models and Air Pollution: Twenty-five Years and Counting. *Environmental and Resource Economics*, 3, 381-394.

Tiebout, C.M. (1956). A Pure Theory of Local Expenditures. *Journal of Political Economy*, 64, 416-424.

Tullock, G. (1971). Public Expenditures as Public Goods. *Journal of Political Economy*, 79, 913-918.

U.S. Census Bureau. (2001). *Statistical Abstract of the United States, 2001.* U.S. Government Printing Office, Washington, D.C.

U.S. Census Bureau. (2006). *Statistical Abstract of the United States, 2006.* U.S. Government Printing Office, Washington, D.C.

White, H. (1980). A Heteroskedasticity-Consistent Covariance Matrix Estimator and a Direct Test for Heteroskedasticity. Econometrica, 48, 817-838.

Reviewed by James V. Koch, Old Dominion University.

In: Environmentalism: Environmental Strategies... ISBN: 978-1-60741-634-0
Editors: E. Kahraman and A. Baig, pp. 191-209

Chapter 10

THE DEMOGRAPHIC DIMENSION OF CLIMATE CHANGE

Emilio Zagheni*
Department of Demography, University of California, Berkeley, CA, USA

Abstract

In this chapter, the demographic dimension of climate change is discussed. In the first part of the chapter, I analyze the leverage of demographic dynamics on demand of energy-intensive consumption goods and carbon dioxide emissions. Several demographic factors are related to greenhouse gas emissions. Population growth is the most obvious one: the larger the population the larger the amount of emissions, given fairly constant per-capita carbon emissions over time. Population age structure is another important factor: given age-specific consumption profiles of energy-intensive goods, changes in population age structure have a considerable impact on overall levels of consumption and greenhouse gas emissions. Household size affects the extent of economies of scale in energy use from cohabitation: larger households are associated with lower per capita energy consumption. In the second part of the chapter, I discuss the role of the economic structure in amplifying or softening the leverage of demographic dynamics on greenhouse gas emissions. Demographic agents are considered in their role of consumers and producers: the implications of a consumption-based versus a production-based approach to the evaluation of the effect of demographic dynamics on greenhouse gas emissions are explicated. The hypothesis of existence of environmental Kuznets curve is analyzed and its validity in the context of greenhouse gas emissions is discussed. The main point of the chapter is that demographic quantities have an important leverage on greenhouse gas emissions. Their effect is mediated by economic and technological factors. A comprehensive policy approach to address the issue of climate change, that takes into account both demographic and economic dynamics is suggested.

1. Introduction

The 2007 report on climate change of the Intergovernmental Panel on Climate Change (IPCC) states that "warming of the climate system is unequivocal, as is now evident from

*E-mail address: emilioz@demog.berkeley.edu

observations of increases in global average air and ocean temperatures, widespread melting of snow and ice, and rising global average sea level" (IPCC, 2007). Global climate changes may have harsh consequences on human societies. Tackling climate change is a big challenge of our times that requires perspectives from several different fields of study.

The observed increase in average global temperatures is most likely related to the observed increase in anthropogenic greenhouse gas concentrations (IPCC 2007). Demographic trends and changes in economic production are the main factors driving greenhouse gas emissions (see, e.g., ONeill, MacKellar and Lutz, 2001).

In this chapter, I discuss the effect of demographic dynamics on greenhouse gas emissions. I focus on carbon dioxide, which is a gas that, together with water vapor, is responsible for most of the greenhouse effect on the Earth.

In the first part of the chapter, I analyze the leverage of single demographic quantities on demand of energy-intensive consumption goods and carbon dioxide emissions. In particular, I consider the effect of population size, population growth, population age structure and household dynamics.

In the second part of the chapter, I discuss the role of economic structure and technology of production in amplifying or softening the leverage of demographic dynamics on greenhouse gas emissions.

Demographic and economic dynamics are strongly interrelated. In this chapter, I show how the interaction of demographic and economic variables plays an important role in shaping the level of emissions of carbon dioxide.

The major implication of this study is that a holistic policy approach should be considered to tackle the issue of climate change. The effectiveness of environmental policies may be undermined by demographic trends. I show how this has been true for several pollutants and how this could be the case for carbon dioxide emissions. The effect of demographic dynamics and their interaction with the economy should be taken into consideration by policy-makers to effectively address the issue of climate change.

2. The Leverage of Demographic Dynamics on Carbon Emissions

2.1. Population Size and Growth

According to the United Nations[1], the estimated world population of 2005 is about 6.5 billion and 80 percent of it resides in less developed regions. The world population in 2005 is about 2.5 times bigger than it was in 1950 and the most rapid growth has occurred in developing regions. According to the United Nations' projections, there will be about 9 billion people populating the Earth in 2050 and 86 percent of them will live in countries that are now considered less developed.

The spectacular growth of the world population over the past 50 years is the result of the demographic transition, a process which started in the 18th century in Europe and that gradually spread to other parts of the world. In the least developed countries, the demographic transition did not start until the early 1950s, but the process has been much

[1]United Nations World Population Prospects, 2006

faster than in European countries and has led to rapid population growth over the second half of the 20th century.

The demographic transition is a process that usually unfolds together with the modernization of a country. The demographic regime of a country shifts from high fertility and mortality rates to low fertility and mortality rates. The mortality rates are usually the first ones to drop, thanks to improvements in areas such as sanitation and infant mortality. Fertility rates drop as well, but with a certain lag. The result is population growth, fast in the early stages of the transition and rather slow in later stages, until a new equilibrium between fertility and mortality rates is reached (see, for instance, Lee, 2003).

In this section, I am interested in discussing the consequences of changing population size and population growth on carbon dioxide emissions. There are a couple of issues that deserve special consideration: first, we want to know how relevant population size and growth are in explaining carbon dioxide emissions. Second, it is important to understand the role and extent of population policies.

In the history of Demographic thought, a lot has been said about the interaction between populations and their environments, the role of population growth on environmental disruption and the feedback mechanisms from the environment to the population. Nonetheless, it is only during the second half of the twentieth century that population started growing at an unprecedented pace and that concerns about the impact of rising population levels became more and more pressing.

In 1968, Paul Ehrlich published the book *The Population Bomb*, in which he warned, in a Malthusian way, of the dire consequences of rapid population growth. The book became a best-seller and stirred debates focusing on the role of population policies and the environment. In the early 1970s, Ehrlich and colleagues formalized their line of reasoning by suggesting a simple identity known as IPAT equation (Ehrlich and Holdren, 1971, 1972; Commoner, 1972). This equation states that environmental Impact (I) is the product of three factors, Population size (P), Affluence (A) and Technology (T):

$$I = P \times A \times T \tag{1}$$

where I stands for environmental impact, P is population size, A is income per capita (Y/P) and T is environmental impact per unit of economic production (I/Y).

The IPAT identity was proposed as a conceptual scheme to understand environmental impact. Population size and population growth have a central role in explaining environmental disruption within the framework of the IPAT equation. The idea is that Population size acts as a multiplier of levels of affluence and technology in the determination of the environmental impact.

If affluence and technology do not change over time, then an increase in population size translates into a proportional increase of environmental impact. In the context of climate change, this story is in line with the fact that per-capita carbon dioxide emissions have been fairly constant since 1970 on a global scale (Smith, 2008). A larger population would thus mean a proportionally larger amount of carbon dioxide emissions.

The argument that more people means a larger amount of carbon dioxide emissions is certainly simplistic and the IPAT equation has several flaws. However, in my opinion, one important merit of IPAT and, more generally, IPAT-based approaches, is that they explicitly consider the population as an important factor that affects environmental impact and, in the context of climate change, carbon dioxide emissions. This means that population policies should be taken into consideration as an option to fight climate change and to make adaptation to climate change easier. Reducing unintended births could be an important step towards rapidly reducing the growth of world population and thus carbon dioxide emissions.

Although it is important to consider population policies, it is also true that reducing unintended births may not be enough or effective to tackle climate change. First, a reduction of unintended pregnancies may not diminish the number of children a woman would have over her entire life. It may only affect the timing of births. Second, even if population growth slows down, and population size stabilizes, carbon dioxide emissions may not reduce proportionately.

There are several reasons why the IPAT identity may be too simple as a conceptual scheme. First, the variables on the right-hand side of the equation are considered independent in the IPAT framework, whereas there are important correlations between them, as Preston (1996) pointed out. Population tends to grow at a faster pace where per capita income is lower. A reduction in global population growth may thus be driven mainly by reductions in population growth in low-income countries. As a consequence, reductions in population growth may result in a less than proportional lessening of carbon dioxide emissions. In addition to that, changes in population growth rates may have an impact on economic growth and technological progress, thus making the relationship between demography and carbon dioxide emissions nonlinear. It is also important to notice that, if we disaggregate our analysis for countries with different per-capita income levels, we observe that in the richest countries the population has been growing slowly, but emissions have increased decidedly. In the poorest countries, population has grown fast, but emissions have not kept the same pace.

The limitations that I just discussed with regard to the analysis of the effect of population on carbon dioxide emissions, within the IPAT framework, are apparent also when we try to analyze the effect of changes in affluence and technology on carbon dioxide emissions. Reduction in per-capita income does not seem to be a politically viable solution to the problem of climate change. However, there is a lot of attention on the role of technology. Decarbornization of the economy seems a solution that could complement population policies and that would make it possible for growing populations to reduce their carbon dioxide emissions.

The IPAT equation has been used extensively in the field of population and climate change. It has been appropriately modified to quantitatively estimate, for instance, the impact of changing demographic, economic and technological factors on carbon dioxide emissions (Dietz and Rosa, 1994, 1997; Rosa and York, 2002). The IPAT scheme has been the intellectual father of lots of interesting works and analyses carried out in the field of population and environment. I think that the most important legacy of the approach is the creation of a framework for thinking of environmental problems in terms of their demographic, economic and technological causes. Reduction of population growth and decarbonization of the economy are the two main policy-relevant suggestions that emerge from an IPAT-based conceptual scheme.

I believe that the IPAT equation is a good starting point for the evaluation of the effect of population size and growth on environmental impact. But there are several other demographic factors, in addition to population size and growth, that affect carbon dioxide emissions and should be taken into account. Some of these factors are closely interrelated with the spread of innovations. For instance, the development of a new transportation system, based on a fuel which is cleaner than oil, may require certain levels of population density in order to make the distribution of the fuel economically viable. Some other demographic processes, such as population aging and changing household structures, are relevant because they have an important impact on consumption patterns of energy-intensive goods.

The point that I would like to make in this chapter is that demographic dynamics matter for carbon dioxide emissions. Population size and growth are only two aspects of a larger set of demographic factors that have an effect on carbon dioxide emissions. The interaction of demographic dynamics with economic and technological variables is the key factor in determining carbon dioxide emissions. If we want to isolate the impact of population growth, then we can say, in a rather naive way, that more people require more energy and thus produce more emissions. This reasoning could be a relatively good approximation to the truth in certain settings, but the role of population growth on carbon dioxide emissions can be fully understood only when a larger set of demographic factors are taken into consideration. In the following sections, I will develop on this line of thought.

2.2. Population Age Structure

Population age structure is an important determinant of overall levels of consumption and production of an economy (e.g., Cutler, Poterba, Sheiner and Summers, 1990; Diamond, 1965; Lee, 1994; Willis, 1988). Therefore, changes in population age structure strongly affect the amount of carbon dioxide emissions of a society. In this section, I outline the mechanisms through which age structure influences carbon emissions throughout the demographic transition.

In the early stages of the demographic transition, infant and child mortality decreases and the proportion of children in the population increases. Later, when the surviving children reach adulthood, the percentage of population in the working ages increases. Improvements in mortality conditions are followed by reductions in fertility levels, thus leading to aging in the final stages of the transition. The proportion of working age population declines, whereas the proportion of elderly increases (e.g. Lee, 2003).

A higher share of working-age population has a positive impact on per-capita income. This is the so-called 'first demographic dividend' that countries experience during the demographic transition. When the population ages, the proportion of working-age people returns to levels similar to the ones observed in the early stages of the demographic transition. However, there is a major difference compared to the situation in the early stage of the transition. In the final stages, the proportion of elderly is much larger than at the beginning of the transition.

Population aging has important macro-economic consequences. Since there are proportionately less working-age people to support an increasingly larger proportion of elderly, per-capita income is expected to decrease. However, this may or may not be true depending on the extent of what is known as 'second demographic dividend'. Population aging may lead workers to accumulate savings to fund their retirement. If that is the case, wealth is mostly accumulated in the form of capital and thus population aging is associated with a higher ratio of capital per worker in the economy. This situation may lead to higher productivity, wages and per-capita income. The extent to which capital accumulation counteracts the unfavorable age distribution, in terms of per-capita income, depends on the relative importance of intergenerational transfers, with respect to individual savings, as a form to fund retirement (e.g. Mason, Lee and Lee, 2008).

What is most important for our purposes is the level of sustainable consumption over the demographic transition. Analogously to per-capita income, when the proportion of working-age people is large, then high level of consumption can be sustained. With population aging, the level of consumption may increase or decrease, depending on the extent of the second demographic dividend and the form of funding for retirees.

Global levels of consumption are strongly related to the age structure of a population. Population aging may increase the level of capital per worker and consumption (e.g. Cutler et al., 1990; Lee, 1994; Mason et al., 2008). Higher consumption would in turn generate more carbon dioxide emissions, all other things being equal. Changing age structure is also related to the age-speicific consumption profiles of energy-intensive goods. Such profiles may be different from the aggregate consumption profile. Zagheni (2008) estimated the consumption pattern by age for a set of consumption goods in the United States. Such estimates show that consumption of energy-intensive goods generally increases with age until the adult stage of life. Then the age-specific consumption tends to either increase or decrease for the elderly, depending on the specific good considered. Zagheni (2008) shows that, if the per-capita consumption profile of energy-intensive goods is rather stable over time, then population aging has a positive noticeable impact on carbon dioxide emissions.

The effect of population aging on carbon dioxide emissions is not completely clear, especially because it is hard to isolate the demographic component of population aging from related socio-demographic processes. Dalton, O'Neill, Prskawetz, Jiang and Pitkin (2008) show that an energy-economic growth model with multiple dynasties of heterogenous households implies that population aging may reduce long term emissions. The uncertainty about the impact of aging on carbon emissions is related to the fact that there is still uncertainty about the role of population aging on economic growth and per-capita consumption. Outcomes of economic models are strongly dependent on the assumptions made, and the effect of population aging is related to institutional factors, such as intergenerational transfers, which vary largely across countries.

It is hard to isolate the effect of aging on carbon emissions. First, population aging may be driven by a decrease in fertility or an increase in life expectancy, or a combination of both. The consequences of population aging may be different depending on which one is the driving force behind it. In addition to that, population aging is usually accompanied by a reduction in household size, which means an increase in household numbers and larger carbon dioxide emissions. Although population aging does not necessarily imply changes in household structures, it is likely to be negatively correlated with household size, through reduced fertility or increased longevity. I therefore believe that population aging may have a positive impact on carbon dioxide emissions, for reasons that I mentioned in this section, and because of households dynamics associated to population aging that I will thoroughly discuss in the next section.

2.3. Household Dynamics

In the previous sections, the individual has been used as unit of analysis to discuss the effect of demographic dynamics on carbon dioxide emissions. In this section, I would like to analyze the role of changing household size and composition on energy use and carbon emissions. The choice of the unit of analysis is particularly important in this context and may significantly affect the results of any decomposition exercise.

In the classical IPAT model, the demographic unit is the individual. This is a reasonable choice if household formation and size are relatively stable over time. But it is not the only possible choice. Households or any other aggregate of individuals, such as a community, could be considered emitting units as well. The choice of households instead of individuals as unit of analysis may be justified by the fact that, for many energy-intensive goods, the household is the unit of consumption and there are significant economies of scale in consumption from cohabitation. Thus, for certain environmental issues, such as emissions of carbon dioxide, the population size and growth rate are of less concern than the number of households and their growth rate (Mackellar, Lutz, Prinz and Goujon, 1995).

When the demographic unit of analysis is the household, instead of the individual, the IPAT identity can be converted into a IHAT identity:

$$I = H \times A \times T \tag{2}$$

where I is carbon dioxide emissions for our purposes, H is the number of households, A is Gross Domestic Product per household and T is the amount of emissions per unit of Gross Domestic Product.

The choice of IHAT as a conceptual scheme is motivated by observed trends in household dynamics. In particular, growth rate of number of households is an important quantity. This rate is associated to the average household size, to the roles of changing age-specific household headship rates and to changing population age structure (Mackellar et al., 1995). More specifically, average household size is the inverse of the average household headship rate, which is defined as the number of people in the population who are heads of household, divided by the population and thus expressed as a proportion. Age-specific household headship rates are defined by the number of heads of household in a certain age group, divided by the total number of people in that age group. Changes in average household size result from the combined evolution of age-specific household headship rates and the age structure of the population.

Between 1950 and 1990, average household size in the more developed regions of the world declined, on average, from 3.6 to 2.7 (Mackellar et al., 1995). This decline was due to population aging and to increases in age-specific household headship rates that reflect changes in behavior (e.g. declining age-specific nuptiality rates, growing tendency of elderly to live on their own and of young adults to move away from home earlier). The shift towards more atomized living arrangements in developed countries is expected to continue in the future and it may generate a larger amount of emissions than it was expected on the basis of projections that rely on individuals as unit of analysis.

Mackellar et al. (1995) analyzed the demographic forces influencing energy consumption and CO_2 emissions within the framework of IPAT and IHAT equations. They found that, on the basis of the IPAT scheme, roughly one-third of the growth of energy consumption in both less developed and more developed regions over the period 1950-1970 was accounted for by demographic growth. Shifting to an IHAT framework does not greatly alter the results of the decomposition analysis for less developed regions, but it completely changes the picture for more developed countries. In the richest countries of the world, about three-quarters of the growth rate of energy consumption is accounted for by demographic increase, as measured by the growth rate of the number of households.

Mackellar et al. (1995) also consider projections of carbon dioxide emissions for the period 1990-2100. If the IPAT approach is used for the decomposition exercise, and emissions are assumed to remain constant in per-capita terms over time, then it is projected that world emissions increase by 88% between 1990 and 2100. If, by contrast, the IHAT approach is used, and emissions are assumed to remain constant in per-household terms, the projected increase is 167%.

The main reason for the rapid growth in number of households is the reduction in household size. This trend is caused by several factors such as lower fertility rates, higher per-capita income, higher divorce rates, population aging, and the decreased frequency of co-habitation of multi-generational families (e.g., Liu, Daily, Ehrlich and Luck, 2003). The evolution of some of these demographic quantities over time is associated to a reduction in population growth and an increase in the growth rate of number of households. For instance, fertility decline brings about deceleration in the overall population growth rate, thus relieving pressure on the environment. At the same time, it also contributes to the formation of smaller households, and to population aging, which may have a negative effect on the environment. The proportion of the population in age groups characterized by high age-specific household headship rates would increase, thus generating more households, which are also smaller and benefit less from economies of scale from cohabitation.

The IPAT and IHAT models represent somewhat extreme formulations, the first assuming no household-level economies of scale in the activities that generate CO_2, and the second ignoring individuals and considering only households. In order to balance the importance of household-level economies of scale from consumption, and of the overall number of individuals in the households, I propose the 'equivalent adult' as unit of analysis. For example, I suggest to use data from expenditure surveys in order to adjust the dimension of households in terms of their environmental impact, using equivalence scales such as:

$$m_h = (s_a + \eta s_c)^\theta \tag{3}$$

with $\eta \in [0, 1]$ and $\theta \in [0, 1]$

m_h is the factor of equivalence related to the household h; s_a and s_c are, respectively, the number of adults and children in the household; η is a parameter that represents the importance of the different needs. $\eta = 0$ means that only adults are considered responsible for environmental impact, whereas $\eta = 1$ means that no distinction is made between adults and children. θ is a parameter that is related to the improved efficiency of the household, compared to the sum of single individuals. $\theta = 0$ is associated to the situation of maximum efficiency. In fact, in this case the number of equivalent adults of the household is 1, thus meaning that a household composed of several members is given the same weight, in terms of environmental impact, as a single adult. On the contrary, $\theta = 1$ means that no efficiency gains arise from living together.

The IHAT formulation makes it clear that the assessment of the driving forces behind environmental impact should take into account not only the population size, but also the role of households and the age structure of the population. What I suggest is a synthesis of IPAT and IHAT approaches: I propose a conceptual framework in which the relevant variable that affects the environment is the number of equivalent adults, so that we can rewrite the IPAT and IHAT identities as IEAT:

$$I = E\frac{Y}{E}\frac{I}{Y} \tag{4}$$

where I represents the environmental impact in terms of carbon dioxide emissions, E is the number of equivalent adults, A may be seen as the Gross Domestic Product per equivalent adult and T is the impact per unit of Gross Domestic Product.

3. The Role of Economic Structure

3.1. Consumers vs Producers

In the previous sections, I mentioned that the leverage of demographic dynamics on carbon emissions is strongly related to economic factors, such as consumption levels, saving rates, per-capita income, etc. In this section, I discuss the role played by people, as consumers or as producers, in shaping the level of carbon dioxide emissions. It is important to distinguish between consumers and producers, to correctly identify the relative importance of production and consumption on carbon dioxide emissions in different contexts and geographic regions.

In order to assess potential actions and policy measures to reduce environmental impact, Waggoner and Ausubel (2002) reconceptualized the IPAT identity, renaming it ImPACT. They disaggregated the IPAT factor T into consumption per unit of GDP (C) and impact per unit of consumption (T), so that the identity becomes:

$$I = PACT = P\frac{Y}{P}\frac{C}{Y}\frac{I}{C} \tag{5}$$

The main purpose of the ImPACT model is to identify the factors that can be altered to reduce pressure on the environment, and the key variables that influence each factor. The identity begins with the environmental impact I, on the left side, whose dimension must be identical to the product of the dimensions on the right side. Waggoner and Ausubel (2002) used emissions of polluting substances as an example of I. In their identity, P is the population size, a variable that is influenced by 'parents'. A stands for affluence and it is defined as the per capita GDP of the geographical region under discussion; it represents the size of the economy and, within the ImPACT scheme, it is affected by 'workers'. C is the intensity of use, such as energy per GDP. 'Consumers' lever C as they decide to employ more or less of their economic muscle on the product that will eventually have an impact on the environment. T is the ratio of environmental impact to goods demanded by people and produced: engineers and industry, the 'producers', can affect this force by inventing, perfecting and employing technology.

The model is a useful tool to understand the extent of each actor's leverage on lessening environmental impact. In fact, the model separates the economic dimension from the choices of consumers and the lever of producers. The effects of lifestyles are separated from those of technologies in lowering impact per economic output.

ImPACT shows that changing environmental impact means changing four multiplying forces: the number of people, the economic muscle of each, the fraction of economic activity devoted to a specific good and the impact of making the good. The phenomenon of dematerialization of the economy is described in terms of declining C. Efficiency improvements are reflected in declining T.

ImPACT assigns the responsibility for environmental disruption to the ultimate actors, who are 'parents', 'workers', 'consumers' and 'producers'. Recently, Zagheni (2008) showed that the IPAT and ImPACT schemes can be generalized within the framework of input-output economic models (Leontief 1953, 1970, 1986; Hendrickson, Lave and Matthews, 2006) in order to differentiate the consumption according to the sectors of the economy to which we can impute the production of the final goods.

Ultimately, there are people broken down by age, sex and other characteristics. These people make economic choices. For instance, they decide which consumption goods to buy. The amount of overall consumption is strongly dependent on population size, age-structure and level of affluence. The leverage of total consumption on carbon emissions is mediated by the existing technology, which is reflected in the flows of resources between sectors.

3.2. Environmental Kuznets Curves

In this section, I consider the impact of economic growth on carbon dioxide emissions. As I introduced in the previous section, the leverage of demography on carbon emissions is strongly affected by the structure of the economy, and the technology in use. In the literature, there is a fair amount of empirical work on the relationship that is assumed to trace the path of environmental impact followed by economies as their per capita GDP grows. Such relationship is important in mediating the effect of demographic dynamics on carbon emissions and will thus be discussed in this section.

One interesting hypothesis, known as environmental Kuznets curve (EKC), is that the relationship between per capita income and the use of natural resources and emission of wastes has an inverted-U shape. The idea behind the EKC hypothesis is that many pollutant levels may rise with GDP at low incomes, pass through a maximum at some characteristic income level and then fall as GDP continues to increase. Some of the reasons for such a path are changes in composition of production and consumption, preferences for environmental quality, and increasing returns to scale associated with pollution abatement

The idea that national environmental quality changes with economic growth may be found in some of the theoretical literature of the 1970s on pollution and growth, where optimal pollution control models have inverted-U curves of pollution implicitly embedded in them (Selden and Song, 1994). In addition, the EKC hypothesis is consistent with the theory of stages proposed by Rostow (1960), who suggested that developing countries are expected to go through structural transformations while moving from traditional to industrial economies and then to a mature mass consumption economy.

Waggoner and Ausubel (2002) show that, within the context of the ImPACT identity, dematerialization and the possibility of environmental Kuznets curves can be regarded as evidence of connections with changing elasticities between income and other forces. Since declining intensity of use, that is decreasing values for C, means dematerialization, raising income dematerializes whenever consumption increases less then proportionately. For example, a curve of intensity of use C first rises as income A grows, but then falls, tracing an inverted-U line related to a first materialization followed by dematerialization. If C is independent of A, no Kuznets curve will be traced; on the other hand, if C is proportional to $A^{(b-1)}$, $\ln(C)$ will rise when b is greater than 1 and dematerialise when b is less than 1. Thus, the changing slope of the inverted-U curve can be regarded as a change from income elasticity to inelasticity as income rises.

The EKC hypothesis has stimulated considerable discussion within and between communities of economists and environmentalists, and debate continues over the validity and the policy implications of EKC studies. According to a production-based approach, the relationship between per capita income and environmental quality depends on scale, composition and technology effects. If the pollution intensity of aggregate output is fairly constant over time, environmental impact is expected to increase with economic growth. As a matter of fact, only the scale effect is involved and greater output generates more pollution. On the other hand, environmental quality could improve with income growth if the scale effect were eclipsed by the combination of composition and technology effects. With increasing per capita income, the composition of output shifts among sectors which differ in their pollution intensity of output. For instance, the service sector may grow relative to the manufacturing sector and this composition effect can alter the pollution intensity of output. Furthermore, the various sectors of the economy may adopt less polluting technologies, either because of market-driven technological advance, stimulated in part by the internal benefits of resource conservation, or government regulation, including standards, taxes and the creation of tradable emissions permits.

Environmental quality may worsen with income within some ranges of GDP, but improve over others. The relationship between impact and GDP may vary according to the dimension of environmental impact taken into consideration. It is plausible, for example, that government regulatory responses are stronger when the effects of pollution lie primarily within national borders than when the impact is primarily transnational or global (Arrow, Bolin, Costanza et al., 1995; Max-Neef, 1995). This suggests that the observation of an environmental Kuznets curve for carbon dioxide emissions is less likely than for some other pollutants, unless international agreements are in place.

Empirical evidence has largely supported the EKC hypothesis and we can recall, as an example of influential contribution, the work of Grossman and Krueger (1995). They found evidence that the relationship between per capita income and environmental degradation takes an inverted U shaped form for 12 of the 14 air and water quality variables considered in their study. An incautious policy inference might propose economic growth itself as a remedy to environmental problems; on the contrary, Grossman and Krueger cautioned against such a reading of their findings arguing that there is no reason to expect that the path followed in the past will be automatically repeated in the future. According to them, the strongest link between income and pollution is provided by an induced policy response, driven by citizens' demand, that takes the form of more stringent and more strictly enforced environmental standards.

A cautionary note on the evidence of EKC is necessary, especially with regard to high-income countries. The possibility that a downturn in pollution, at the peak of the inverted-U curve, is followed by a later upturn that marks a reversal of tendency has been tested in the 'Kuznetsian' literature by using a cubic functional form for income. Pezzey (1989), for example, has argued that an inverted-U relationship may not hold in the long run and he foresees a so-called N-shaped curve which exhibits the same pattern as the inverted-U curve initially, but beyond a certain income level the relationship between environmental pressure and income becomes positive again. The negative correlation is thus considered a temporary phenomenon: once technological improvements in resource use or abatement opportunities have been exhausted or have become too expensive, further income growth will result in net environmental degradation.

Torras and Boyce (1998) argue that while the industrial composition effect, which accompanies rising per capita income, may lower the marginal pollution intensity of output, it cannot fully offset the scale effect unless more pollution-intensive sectors shrink absolutely. This could happen only if these sectors produce inferior goods, whose consumption falls with rising income, or if their products were replaced by imports. In general, the former condition seems unlikely to hold, while the latter simply states a relocation of pollution to other countries.

The hypothesis of a relocation of the environmental impact among countries leads to the consideration of a consumption-based approach to explain environmental degradation. The IPAT and, more specifically, the ImPACT formulations are models consistent with the consumption-based approach. They allow for the presence of variables such as per capita consumption and the intensity of use, that is, the ratio between the consumption of products that eventually impact the environment and GDP.

The parallels between scale, composition and technique on the one hand, and population, consumption and technology on the other hand, are fairly obvious and may support the argument that either a production or consumption based approach would be equivalent for examining the EKC hypothesis. But a problem that arises when adopting a production-based approach is that most environmental degradation can be traced back to the behaviour of consumers, either directly or indirectly: goods and services will not be produced, bought, sold and traded across borders, unless there is a demand for them (Duchin, 1998).

Ekins (1997) raises reservations about using production-based approaches when a significant share of the changes in environmental impact are attributed to changing composition. If the shift in production patterns has not been accompanied by a shift in consumption patterns, environmental effects due to the composition effect are only displaced from one country to another, rather than reduced. Moreover, it is also possible that a reduction in environmental impact will not be available to the latest developing countries, because there will be no countries coming up behind them to which environmentally-intensive activities can be relocated.

A look at data on consumption trends may provide some insights about changes in the composition of consumption. Rothman (1998) deals with the question of changes in the composition of consumption on the basis of cross-sectional data. He presents data on the quantities of per capita consumption for eight categories of consumer goods for a large number of countries in 1985, and he relates these data to the income level of the countries. Nearly all categories of consumption display growth with income. Therefore a changed composition, in terms of shares, is due to differences in relative growth rates. The only category exhibiting a fall with high levels of income is food. But when this category is broken down into three groups (grain and starches, meat and animal products and other foods), the fall is concentrated on grains and starches, the food items that have relatively low input intensities. Rothman combines his own observations with references to other studies and concludes that virtually no evidence exists that a change in the composition of consumption has resulted in decreasing environmental impact at higher levels of income.

Ropke (2001) examines the hypothesis of environmental Kuznets curve with a consumption based approach, thus decomposing environmental impact into three determinants: the level of consumption, the composition of consumption and the environmental intensity of the goods and services produced for consumption, including both direct and indirect effects. In his work, he conceptualizes environmental impact as related to three types of physical input (material, land and energy) and he defines the environmental impact of a specific consumption good as the appropriation of inputs necessary to provide a unit of that good, including not only the direct use of the inputs in the production of the good, but also the indirect use of inputs that have been employed for the provision of raw materials, intermediate products and capital equipment. He calls *input intensity* of a given consumption good the quantity of a given input used for the provision of either one unit of the good or of the quantity of the good that can be bought for a specific monetary amount. Ropke's analysis supports Rothman's point that consumption of nearly all categories of goods and services increases, while the rates of increase differ. A common trend in rich countries is that the share of food consumption falls, whereas the housing and transportation shares increase. Ropke, in particular, analyzes the Danish case, showing that the main shifts in the composition of consumption have meant that some categories of consumption with a high environmental impact have replaced other categories with a high impact. Therefore, the changing composition has not contributed to making consumption more environmentally-friendly.

The possibility for a disconnection between national patterns of production and consumption is made possible by international trade, which has therefore become a major issue of consideration in examining the relationship between economic growth and environmental impact. Much has been written in recent years on the relationship between trade, trade liberalization, environmental quality, environmental policy and economic performances, but there is not a wide consensus in the academic world. To our purposes, it is interesting to assess to what extent changes in resource use and environmental degradation with income level are due to shifting pollution and resource intensive production across borders.

In the context of global climate change, for example, proposals for limiting CO_2 emissions often call for high income countries to take the lead in emissions reductions by imposing carbon taxes. However, there is a strong possibility that energy intensive industries would relocate from countries that impose a high carbon tax to those which do not. This relocation could counteract any efforts at reducing global emissions, especially if the industries shift to countries that use energy less efficiently.

It is not sufficient to examine the pollution of energy intensive industries in order to capture the full impact of international trade on the environment, since there are a number of final manufactured goods whose production only releases a small amount of pollutants directly, but has an indirect impact because pollution-intensive inputs are used in the production chain. This point is considered by Wyckoff and Roop (1994) who estimate that nearly 50% of the carbon embodied in the imports of major developed countries was embodied in non energy intensive goods.

In order to capture both the direct and indirect effects of trade on pollutant generation,

Suri and Chapman (1998) developed a model in which they introduced several explanatory variables for environmental impact whose dynamics are significant in the context of an analysis of cross-country movement of goods that embody pollution.

A first variable they include is the ratio of imports of all manufactured goods (both energy and non energy intensive) to domestic production of all manufacturers (the *import-manufacturing ratio*). In the 1971-1990 period this ratio increased in nearly all industrialized countries, while in most developing countries this ratio remained steady or declined and in countries where it rose, the rate was much lower than in the industrialized nations. To some extent, the behaviour of this ratio in developing countries reflects import substituting industrialisation.

A rise in the import-manufacturing ratio can have opposite effects on energy requirements. On the one hand, imports could substitute some manufactured goods that were earlier produced domestically. This would include those industries that have developed overseas in response to stricter environmental standards in the home country. Also included here would be industries that have been attracted to other countries with lower labour and raw material costs. Another case would be the introduction of a new production process that replaces domestically produced inputs with imported ones. Ceteris paribus, imports of commodities, by replacing domestic production, may reduce the energy requirements of a country. On the other hand, imports of certain types of machinery could lead to an increase in energy requirements. This would happen in the case of capital goods which are used to produce goods that were not manufactured domestically before and whose energy intensity is higher than the average for other domestic production. It is likely that countries that are industrialising and hence expanding their production base, would be the ones that have larger imports of such capital goods than the industrialized ones. Therefore, in these low and middle income countries the positive effect on energy requirements can be quite large; on the contrary, in the high income countries imports are supposed to have a strong substitution effect on energy demand.

A second interesting variable included in the analysis of Suri and Chapman (1998) is the ratio of exports of manufactured goods to domestic manufacturing production (the *export-manufacturing ratio*). The increase in energy requirements as a result of a rise of this ratio represents the impact of the country's participation in international trade: the embodied energy that is required to produce exports is 'consumed' in another country. Although the ratio has been rising in nearly all the countries in the sample they considered, the rate of growth has been much higher in the industrialising ones. This growth represents the exploitation of overseas markets by these countries in their drive towards industrial maturity.

A third variable, that also reflects the level of industrial maturity of a country is the ratio of total manufacturing value added to GDP: the share of total manufacturing in GDP includes production of manufactures for the domestic market in addition to production for export and it is a measure of the broad structural transformation that an economy goes through as it develops. Evidence suggests that while the overall structural transformation of the economy is leading the industrialising countries towards greater energy intensity, the richer nations are generally moving in the opposite direction.

Suri and Chapman (1998), in their analysis of the impact of international trade on commercial energy consumption, show that imports play an important role in causing the EKC downturn in developed countries. The introduction of trade variables, in models built for detecting the presence of EKC, raises the turning point of the curve for energy consumption; other things being equal, this would imply that incorporating trade effects would also tend to increase the turning point for emissions related to energy use.

In rapidly industrialising countries exports of manufactured goods have been a key source driving their energy consumption. This rapid increase in energy requirement, combined with negligible controls on pollutants, has thus resulted in a rapid increase in emissions of polluting substances.

Industrialized countries, on the other hand, have benefited from the relocation of industrial production, that helped them to reach the flatter part of the EKC for energy consumption: the percentage increase in energy requirements associated with a percentage increase in GDP is now substantially lower in these countries than it was in their industrialisation phase.

In this section, I showed that consumption patterns have an important role in determining the level of energy requirements and thus carbon dioxide emissions. There may be deviations of consumption patterns from production patterns, made possible by international trade and inequalities among countries. A simplistic view of the relationship between income and carbon emissions may lead us to think that as income grows carbon emissions will decrease, thanks to improvements in efficiency. This may or may not be true. In this section, I explained some mechanisms through which income is associated to energy requirements and carbon emissions. The role of international trade and the relative size of different sectors in the economy may account for large part of the variation in national accounts of carbon emissions. I therefore recommend a consumption-based approach to estimate the impact of demographic dynamics on carbon dioxide emissions.

4. Conclusion

In this chapter, I discussed the interaction of demographic and economic dynamics and their effect on greenhouse gas emissions, in particular carbon dioxide. I analyzed the leverage of single demographic quantities on demand of energy-intensive consumption goods and carbon dioxide emissions. I then considered the role of economic structure and technology of production in amplifying or softening the leverage of demographic dynamics on greenhouse gas emissions.

I showed that decarbonization of the economy is certainly an appealing solution to the problem of reducing carbon emissions. However, demographic dynamics may undermine the effectiveness of some environmental policy measures and should thus be taken into account by policy makers.

Population policies should be considered as an option and an opportunity to reduce carbon dioxide emissions. Reducing unintended births may not be an effective solution to the problem of climate change. However, it may help in alleviating the problem and it may have several other positive social consequences, other than the environmental ones.

Policies that encourage environmentally-friendly behaviors may be particularly interesting. For instance, incentives to share electric appliances between more than one household may help improve efficiencies in consumption and counteract the effects of the demographic trend towards atomized households. Most individual choices may have little leverage by themselves, but when we aggregate over a population, their impact becomes significant.

I showed that the same individual choice may have fairly different implications, depending on the geographic location of the individual and the production units. The same consumption good may have different quantities of carbon emissions embedded into it, depending on the country in which it is produced, and therefore the technology that is used. From a production-based viewpoint, delocalization of production has important effects on national accounts of emissions. From a consumption-based viewpoint, migration and assimilation of life styles has a relevant impact on carbon dioxide emissions.

Demographic events, such as births, household formation, household dissolution and migration, have noticeable effects on carbon emissions, when aggregated at the population level. There is a mismatch between those people who are mostly responsible for climate change and the ones that suffer the consequences the most. The individual who makes economic or demographic choices does not necessarily experience the consequences of its actions, thus leading to problems of externalities (Hardin, 1968; Lee, 1990). Although it may be unfeasible to have environmental taxes on giving birth or on union dissolution, it is important to remember that demographic trends may generate inequities, which manifest themselves in the distribution of consequences of climate change. An appropriate system of transfers of economic resources should thus be put in place to balance these demographic-driven environmental inequities.

References

[1] Arrow, K., Bolin, B., Costanza, R., Dasgupta, P., Folke, C., Holling, C.S., Jansson, B.O., Levin, S., Mäler, K.G., Perrings, C., & Pimentel, D. (1995). Economic Growth, Carrying Capacity, and the Environment. *Science* 268, 520-521.

[2] Commoner, B. (1972). The Environmental Cost of Economic Growth. In R.G. Ridker (Ed.), *Population, Resources and the Environment* (1972, pp. 339-364), Washington D.C.: Government Printing Office.

[3] Cutler, D., Poterba J., Sheiner L. & Summers L.. (1990). An Aging Society: Opportunity or Challenge?. *Brookings Papers on Economic Activity* 1, 1-73.

[4] Dalton, M., O'Neill B.C., Prskawetz A., Jang L. & Pitkin J. (2008). Population Aging and Future Carbon Emissions in the United States. *Energy Economics* 30, 642-675.

[5] Diamond, P.A. (1965). National Debt in a Neoclassical Growth Model. *American Economic Review* 55, 1126-1150.

[6] Dietz, T. & Rosa, E.A. (1994). Rethinking the Environmental Impacts of Population, Affluence and Technology. *Human Ecology Review* 1, 277-300.

[7] Dietz, T. & Rosa, E.A. (1997). Effects of Population and Affluence on CO_2 Emissions. *Proceedings of the National Academy of Sciences of the USA* 94, 175-179.

[8] Duchin, F. (1998). *Structural Economics: Measuring Changes in Technology, Lifestyles and the Environment*. Island Press, Washington DC.

[9] Ekins, P. (1997). The Kuznets Curve for the Environment and Economic Growth: Examining the Evidence. *Environmental Planning* A 29, 805-830.

[10] Ehrlich, P. (1968). *The Population Bomb*. A Sierra Club-Ballantine Book, N.Y.

[11] Ehrlich, P. & Holdren J. (1971). Impact of Population Growth. *Science* 171, 1212-1217.

[12] Ehrlich, P. & Holdren, J. (1972). Impact of Population Growth. In R.G. Ridker (Ed.), *Population, Resources and the Environment* (1972, pp. 365-378), Washington D.C.: Government Printing Office.

[13] Grossman, G.M. & Krueger, A.B. (1995). Economic Growth and the Environment. *Quarterly Journal of Economics* 112, 353-378.

[14] Hardin, G. (1968). The Tragedy of the Commons. *Science* 162, 1243-1248.

[15] Hendrickson, C.T., Lave L.B. & Matthews H.S. (2006). *Environmental Life Cycle Assessment of Goods and Services, An Input-Output Approach.* Resources for the Future. Washington D.C., USA.

[16] Intergovernmental Panel on Climate Change (IPCC). (2007). Climate Change 2007. Fourth Assessment Report. Un-edited Copy Prepared for COP-13.

[17] Lee, R. 1990. Comment: The Second Tragedy of the Commons. *Population and Development Review* 16, 315-322.

[18] Lee, R. (1994). The Formal Demography of Population Aging, Transfers, and the Economic Life Cycle. In L. Martin and S.H. Preston (Eds.), '*The Demography of Aging* (1994, pp. 8-49), National Academy Press.

[19] Lee, R. (2003). The Demographic Transition: Three Centuries of Fundamental Change. *Journal of Economic Perspectives* 17(4), 167-190.

[20] Leontief, W. (1953). *Studies in the Structure of the American Economy.* Oxford University Press, New York.

[21] Leontief, W. (1970). Environmental Repercussions and the Economic Structure: An Input-output Approach. *Review of Economics and Statistics* (LII) 3, 262-271.

[22] Leontief, W. (1986). *Input-output Economics*. New York: Oxford University Press.

[23] Liu, J., Daily, G.C., Ehrlich, P.R. & Luck, G.W. (2003). Effects of Household Dynamics on Resource Consumption and Biodiversity. *Nature* 421, 530-533.

[24] MacKellar, F.L., Lutz, W., Prinz, C. & Goujon, A. (1995). Population, Households and CO_2 Emissions. *Population and Development Review*. 21(4), 849-865.

[25] Mason, A., Lee S.H. & Lee, R. (2008). Will Demographic Change Undermine Asia's Growth Prospects?. Bacground paper for *Emerging Asian Regionalism: Ten Years after the Crisis. A Study of the Asian Development Bank*.

[26] Max-Neef, M. (1995). Economic Growth and Quality of Life: A Threshold Hypothesis. *Ecological Economics* 15, 115-118.

[27] ONeill, B.C., MacKellar, F.L., & Lutz, W. (2001). *Population and Climate Change*. Cambridge University Press.

[28] Pezzey, J., (1989). Economic Analysis of Sustainable Growth and Sustainable Development. Environment Department Working Paper No 15, The World Bank.

[29] Preston, S.H. (1996). The Effect of Population Growth on Environmental Quality. *Population Research and Policy Review* 15, 95-108.

[30] Ropke, I. (2001). The Environmental Impact of Changing Consumption Patterns: a Survey. *International Journal of Environment and Pollution* 15(2), 127-145.

[31] Rosa, E.A. and York, R. (2002). Internal and External Sources of Environmental Impacts: a Comparative Analysis of the EU with other Nation Groupings, *National Europe Centre Paper No. 22*, Paper presented to conference on *The European Union in International Affairs*, National Europe Centre, Australian National University, 3-4 July 2002.

[32] Rostow, W. (1960). *The Stages of Economic Growth*. Cambridge University Press, Cambridge.

[33] Rothman, D.S. (1998). Environmental Kuznets Curves - Real Progress or Passing the Buck? A Case for Consumption-based Approaches. *Ecological Economics* 25,177-194.

[34] Selden, T. and Song, D. (1994). Environmental Quality and Development: Is there a Kuznets Curve for Air Pollution Emissions? *Journal of Environmental Economics and Management* 27, 147-162.

[35] Smith, K. (2008). The Population Problem. *Nature Reports Climate Change* 2, 72-74.

[36] Waggoner, P.E. and Ausubel, J.H. (2002). A Framework for Sustainability Science: a Renovated IPAT Identity. *Proocedings of the National Academy of Sciences* 99(12), 7860-7865.

[37] Suri, V. and Chapman, D. (1998). Economic Growth, Trade and Energy: Implications for the Environmental Kuznets Curve. *Ecological Economics* 25, 195-208.

[38] Torras, M. and Boyce, J.K. (1998). Income, Inequality, and Pollution: a Reassessment of the Environmental Kuznets Curve. *Ecological Economics* 25, 147-160.

[39] Wyckoff, A.M. and Roop, J.M. (1994). The Embodiment of Carbon in Imports of Manufactured Products: Implications for International Agreements on Greenhouse Gas Emissions, *Energy Policy* March, 187-194.

[40] Willis, R. (1988). Life Cycles, Institutions and Population Growth: A Theory of the Equilibrium Interest Rate in an Overlapping-generations Model. In R. Lee, W.B. Arthur and G. Rodgers (Eds.), *Economics of Changing Age Distributions in Developed Countries* (1988, pp. 106-138), Oxford University Press.

[41] Zagheni, E. (2008). The Impact of Demographic Dynamics on Natural Resources. A Discussion within the Framework of Input-output Models. Paper presented at the *Population Association of America Annual Meeting* , New Orleans, Louisiana, April 18 2008.

INDEX

D

E

F

J

K

L

M

N

O

P

Q

R

V

W

X

Y

Z